New Britain Diary, 1954:

An Anthropologist's Journal

First Edition

Ravenna Press
PO Box 127
Edmonds, WA 98020

ISBN   978-0-9791921-1-1
ISBN   0-9791921-1-0

Library of Congress Control Number: 2007924569

This book is dedicated to the Lakalai people of the
Hoskins Peninsula, New Britain, Southwest Pacific.

Contents

Acknowledgements

Many people have contributed to the writing of this book in so many ways. First I would like to thank Ward Goodenough, Ann Chowning and the late Charles Valentine and his wife Edith, my companions on New Britain. I am most grateful to the following individuals for their kindness and hospitality, Assistant District Officer Michael Foley, Medical Administrator A. V. "Bill" Bell of Talasea, Cynthia Smith of the Malalia Mission, Father Heinrich Berger and the Sisters of the Valoka Mission, and Frank Maynard of the Matavulu Plantation. In particular, I wish to acknowledge the help and friendliness of the late Patrol Officer Ernest Sharp and his wife Mavis.

I am especially indebted to Claudia Gross, Research Fellow in Anthropology at the University of Auckland and Ann Chowning (Professor Emerita at Victoria University of Wellington) for sending me additional photographs for my book.

I gratefully acknowledge the many helpful suggestions that my wife, Kathy Swindler, who has been closely associated with the writing of this book from the beginning, has made. Also, I wish to thank Cooper Renner for his generous contribution of time in reading the manuscript. Dana Swindler deserves special thanks for his help and support in the publication of the diary.

Thanks are due Holger Jebens Research Fellow at the Frobenius-Institut and Managing Editor of Paideuma, Frankfurt am Main, Germany, who contributed the Epilogue. He is editor of *Cargo, Cult, and Culture Critique*, University of Hawai'i Press, 2004, and author of *Pathways to Heaven*, Berghahn Books, 2005.

And, finally, I remember all the help and good times I had together with Kalua and Sege, friends I will never forget.

Prologue

The Expedition

In 1951, the University Museum and the Department of Anthropology, University of Pennsylvania, sent Dr. Ward H. Goodenough, Professor of Anthropology, on a reconnaissance trip to several areas in Papua and New Guinea to decide on an area for future study. He chose New Britain Island, Melanesia, in the Southwest Pacific for this future ethnographic research of the West Nakanai who lived along the Hoskins Peninsula on the north coast of New Britain (the name Nakanai was used in 1954, but since it is Lakalai in their dialect, Lakalai is now the preferred designation for these people.)

In 1954, the expedition under his direction to New Britain was organized. Accompanying Dr. Goodenough were four graduate students: Miss Ann Chowning, Mr Charles A. Valentine and his wife Edith, and myself. The expedition was aided financially by the department of Anthropology of the University of Pennsylvania, the American Philosophical Society and the Tri-Institutional

Pacific Program (a research program jointly administered by the Bishop Museum, University of Hawaii, and Yale University.) Mr. Valentine's work was helped by a Fulbright Scholarship to the Australian National University, and my research was assisted by a pre-doctoral Fellowship from the Wenner-Gren Foundation for Anthropological Research.

Dr. Goodenough, Ann, and the Valentines would study the language and culture, while my investigation would be the physical anthropology of the Lakalai that would include anthropometry (head and body measurement), dental impressions of their teeth, blood samples and fingerprints. Dr. Goodenough, Ann, and I would leave for New Britain in February while the Valentines would leave in April.

I decided to keep a diary of the trip from the time I left Philadelphia in February until I returned in July. This book is a ledger of my several flights across ten Time Zones to and from New Britain and a daily accounting of the five months I spent living with the Lakalai. It is the story of my life with the Lakalai and our many experiences together. My major purpose on the expedition was to collect data for a doctoral dissertation in Anthropology. I received my degree in 1959, and I wish to thank the many wonderful Lakalai who helped me achieve that goal and who enriched my life with their understanding and willingness to participate in my project.

History and Geography of New Britain

New Britain was discovered by Europeans in 1616 when the Dutch Captain Jakob Le Maire sighted it while sailing around New Guinea. He thought it was part of the landmass including New Guinea and New Ireland. The British navigator and buccaneer William Dampier realized it was an island in 1699-1700 and named it New Britain. In 1884, New Britain became Neu-Pommern (New Pomerania), a German Protectorate. It was mandated to Australia following World War I, taken by the Japanese in 1942, and occupied by the America Army in 1945. It subsequently became part of the UN Trust Territory of New Guinea and was administered by Australia. It became part of Papua New Guinea in 1975 when that nation attained independence.

New Britain is located about 5° S latitude and extends from 148° E to nearly 152° E longitude Northeast of New Guinea, in the Territory of Papua and New Guinea in the Southwest Pacific known as Melanesia (Map 1). Map 1 outlines four of the major cultural regions of the Pacific: Polynesia (*Many Islands*) ranging from Easter Island in the East, Hawaii in the Northwest, and New Zealand in the Southwest; Micronesia (*Small Islands*), the Marianas, Carolines, Marshalls, and Gilberts; Melanesia (*Black Islands*), New Guinea, Bismarck Archipelago, Solomons, New Caledonia, and Fiji; Australia.

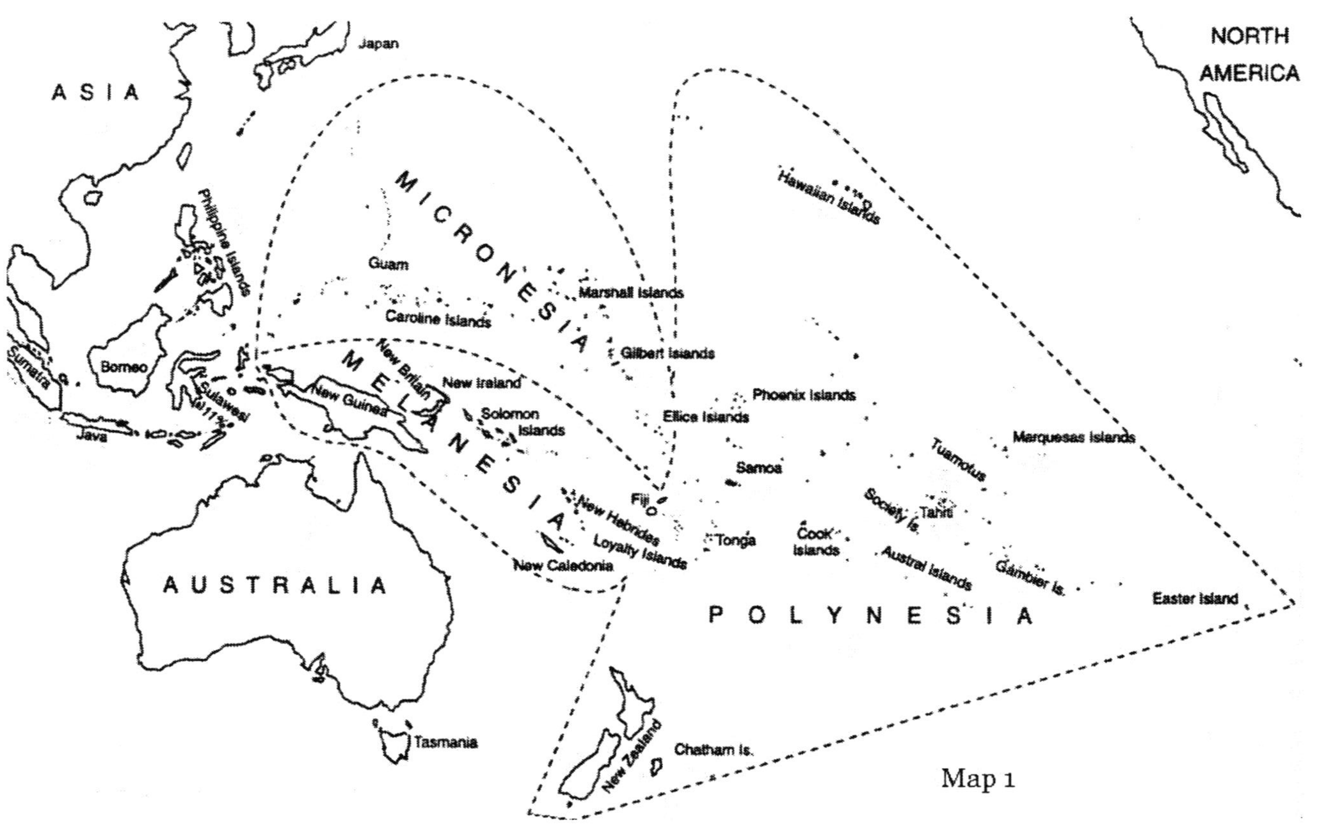

ASIA
Japan
NORTH AMERICA
Sumatra
Borneo
Java
Philippine Islands
Sulawesi
MICRONESIA
Guam
Caroline Islands
Marshall Islands
Gilbert Islands
Hawaiian Islands
Phoenix Islands
MELANESIA
New Britain
New Ireland
New Guinea
Solomon Islands
Ellice Islands
New Hebrides
Loyalty Islands
New Caledonia
Fiji
Samoa
Tonga
Cook Islands
Society Is.
Tahiti
Tuamotus
Marquesas Islands
Austral Islands
Gambier Is.
Easter Island
AUSTRALIA
Tasmania
New Zealand
Chatham Is.
POLYNESIA
Map 1

New Britain is rather crescent-shaped and narrow, about 50 miles wide by 370 miles long, with an irregular coast- line. It is one of several islands making up the Bismarck Archipelago which includes New Island, the Admiralty Islands and an adjacent group of smaller islands (Map 2). New Britain is the second largest island in Melanesia, New Guinea being the largest, and the second largest island in the world.

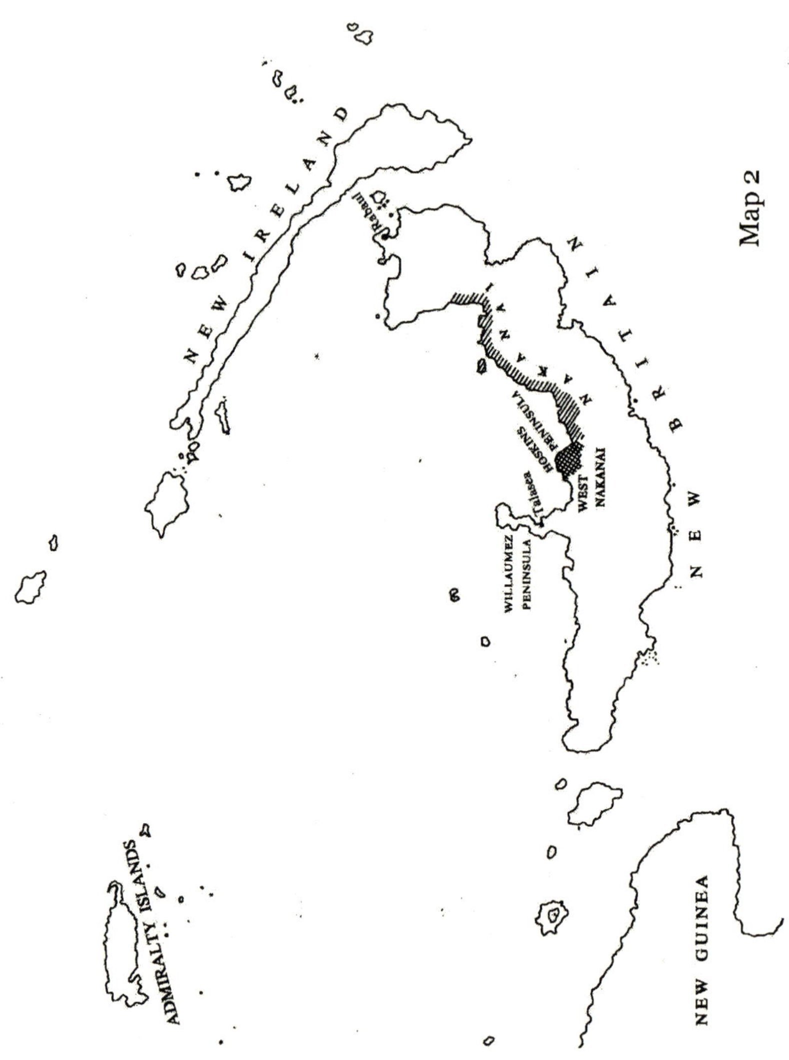

Map 2

Along with several other islands in Melanesia, New Britain represents part of the mountain-building earth movements that rose from the bed of the ocean dating back several tens of millions of years and referred to as oceanic islands. There are many volcanoes along the northern region of New Britain that render this region highly unstable.

A rugged, central system of mountains, composed of the Whiteman, Nakanai, and Baining ranges, extends from one end of the island to the other. Several mountain peaks exceed 7,400 feet in altitude and separate the island into north and south geographical regions. The mountains are geologically young, with many sharp hogbacks (a long, narrow ridge formed from very steep or vertical rock layers), their steep sides forming deep V-shaped valleys. The lowlands are small, scattered, and coastal. A rain forest lies behind the coast and makes up the interior of the island.

Precipitation is high over the island, particularly along the south-central region where it averages about 250 inches a year. The north coast, includingthe Hoskins Peninsula, averages about 151 inches of rain a year mostly during the monsoon season (rain-bearing winds) between December and April and is relatively dry during the rest of the year. During this so-called dry season, the climate is tropical, fluctuating from warm to hot along the coastal regions.

On the Hoskins Peninsula, the hottest hours of the day (between 12 and 3 in the afternoon), temperatures average between the mid 80's and 90's F during the dry season. The evenings and nights are lovely for sleeping since a cool sea-to-land breeze starts about dusk.

The Hoskins Peninsula is a rather flat region between the sea and an arc of rugged mountains that include Mt. Oto at the western end and the active volcano Mt. Pago at the eastern end of the chain (Map 3).

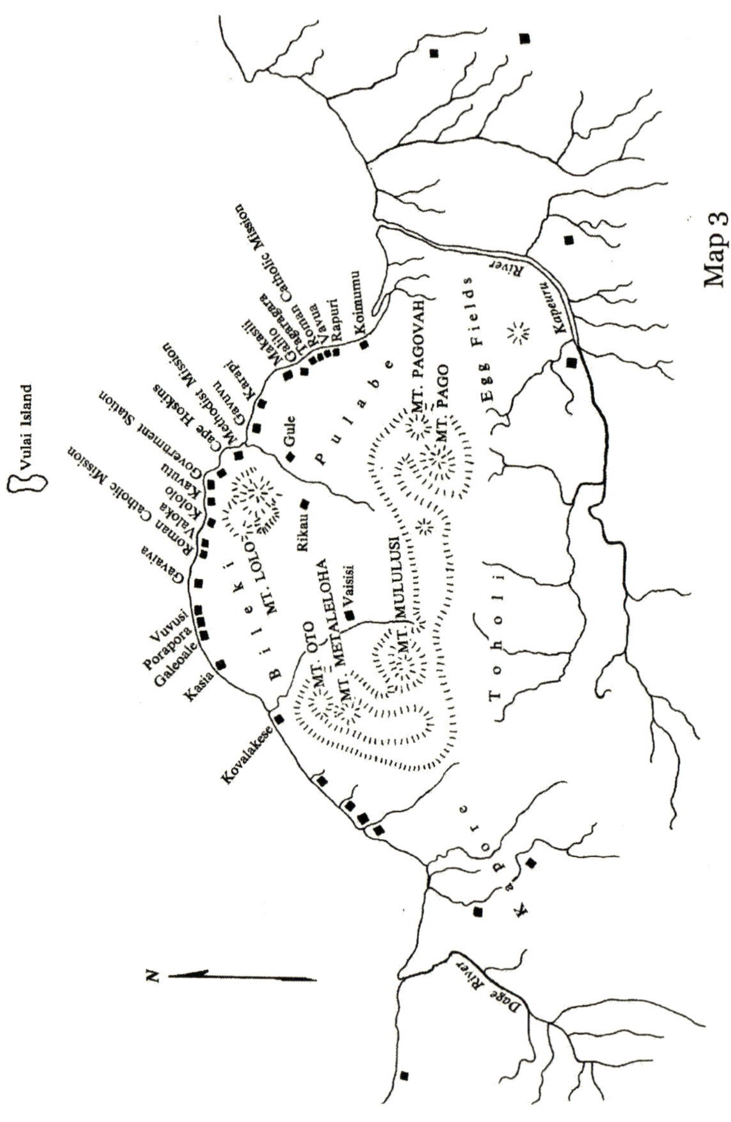

The rain forest of the interior is replaced near the coast by bush and the dense secondary growth of fallow garden land. In this bush the Lakalai have their villages and gardens. The Lakalai, in the time of our expedition, live in nineteen villages, and except for three that are located somewhat inland, are situated near the beach (Map 3). Gardens are the principal source of livelihood and are maintained throughout the year behind the villages in a tract of bush cleared and worked by various combinations of people: for example, a man and woman as gardening partners, a husband and wife, or a brother and sister. The soils are mostly of volcanic origin and can be quite productive, especially those derived from certain basalts. The crops taken from the gardens constitute the major source of food. A variety of cultivated crops are grown, such as taro, *Colocasia sp.*, a member of the Araceae family to which the calla lily belongs, is grown in moist soils. Its thick, starchy root is edible. Its tender leaves are eaten as greens when cooked. Yams, sweet potatoes, sugar cane, tomatoes, to-bacco, cucumbers, bananas, and beans are also grown. Bread-fruit, *Artocarpus sp.*, is a tree that may grow wild, or more often, is cultivated and produces a large fruit that must be cooked or baked before eating. The coconut palm provides supplemental food, drink and building materials. The betel nut palm, *Areca catechu*, and the betel leaf from the betel pepper plant, *Piper betel*, are grown in all villages. A variety of citrus plants are also grown.

Fish and shellfish from the sea and rivers as well as bush-fowl (megapode) eggs provide the major source of protein foods. Megapode eggs are large with rather dull white or cream-like shells containing one of the largest proportions of yolk of any bird. The Melanesian Megapode, *Megapodius eremita*, is a burrow nester. On New Britain the Megapode lays its eggs in burrows of volcanic soil that in Lakalai are directly south east of the coastal villages (Map 3).

Meat is obtained from bush-fowl, wild and domesticated pigs, cuscus, and cassowaries. In New Britain the flightless cassowary bird, *Casuarius sp.*, is smaller than those in New Guinea, and with the emus belongs in the order Casuariiformes. Their eggs are large and green and are incubated by the males in a leafy nest on the ground; and, I might add, they are very good to eat. The cassowary has a dagger-like nail on the innermost of its three toes and can be quite dangerous. The cuscus is a marsupial of the genus *Phalanger sp.* They are about 4 feet long from head to tail, with body and tail approximately the same length. The tail is prehensile and naked for much of its length. This animal spends most of its time in the trees eating leaves and fruit. Use of the meat of these four animals is usually reserved for festive occasions rather than for everyday fare.

A village usually consists of one or more hamlets that contain four to six dwellings and a men's house. Hamlets are traditionally separated from one another by strips of bush that often contain areca palms. Neighboring villages which are connected by paths and which maintain peaceful relations with each other are called territories. Territories are separated by bush or forest and, historically, relations between them were hostile. In 1954, however, they are generally peaceful.

The Lakalai population in 1954 total some 2692 people distributed rather unevenly in nineteen villages along the Hoskins peninsula (Map 3 and Table 1).

Table 1. Population Figures for the Lakalai in August 1954 by Patrol Officer Ernest Sharp

VILLAGE	MEN	WOMEN	CHILDREN	ABSENT	TOTAL
KOIMUMU	41	32	53	4	130
RAPURI	21	21	33	5	80
VAVUA	32	36	84	7	159
GALILO	64	59	125	17	265
MAKASILI	33	21	30	2	86
KARAPI	55	54	119	9	237
GAVUVU	38	35	66	4	143
GULE	20	25	48	12	105
RIKAU	28	40	54	4	126
KAVUTU	21	21	33	4	79
KALOLO	21	18	11	18	68
VALOKA	42	46	53	52	193
GAVAIVA	46	49	71	3	169
VOVOSI	20	25	34	22	101
PORAPORA	37	39	41	36	153
GALEOALE	40	46	69	11	166
KASIA	30	48	92	18	188
KOVALAKESE	37	38	55	17	147
VAISISI	36	21	37	3	97
TOTALS	**662**	**674**	**1108**	**248**	**2692**

There are more than sixty matrilineal sibs (a sib is a unilinear descent group in which actual descent from a common ancestor is not directly traceable.) The sibs are associated with some sacred place or thing such as a mountain, river, animal, or fish. Incidentally, if a sib is named after an animal or fish, it is taboo for the members to eat that particular fish. Sibs with a common food taboo are "sibmates," and these relations extend to all tribes with whom there are connections. In 1954 there are no preferential marriage customs among the Lakalai, except that marriage within the hamlet, village, and territory is preferred over marriage to distant-living people. Marriage to siblings and immediate relatives is forbidden. Kinship considerations are implicit in practically every Lakalai relationship.

The Lakalai

The physical characteristics of the Lakalai in 1954 may be generalized as follows; the average stature for males is 5 feet 5 inches, for females 5 feet 1 inch. Skin color ranges from medium to dark brown. Extremely dark brown or "black" is not present. Head hair is abundant, its color is difficult to observe since it is usually cosmetically altered by dying it black or reddish-orange. When hair is seen in its natural state it ranges from brown to black and can be straight, wavy, quite curly or, when combed to stand upward, it appears to be frizzy. Face and body hair is scarce. The face is generally rather broad. The nose is also broad and possesses a rather high bridge. Neither head length nor breadth is extreme when compared to other populations. Head height however, is rather high, a condition found throughout Melanesia. In most physical characteristics then, the Lakalai fall well within the ranges for Melanesian people for whom there is such formation. Some of the data I intended to collect are standard anthropometric measurements of head height, trunk height, standing/sitting height and other dimensions related to stature and overall body type.

The dress of the Lakalai women consists of a belt and apron of short leaves in front and behind. The leaves are grown in a garden near the village. Women may wear blouses but more often are bare above the grass apron (Fig. A).

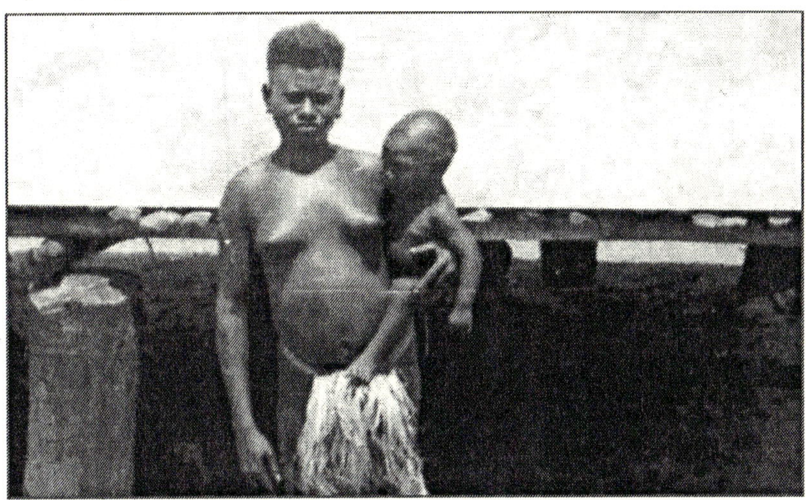

The Lakalai men wear a corresponding cloth wrap-around or *laplap* and are bare above (Fig. B).

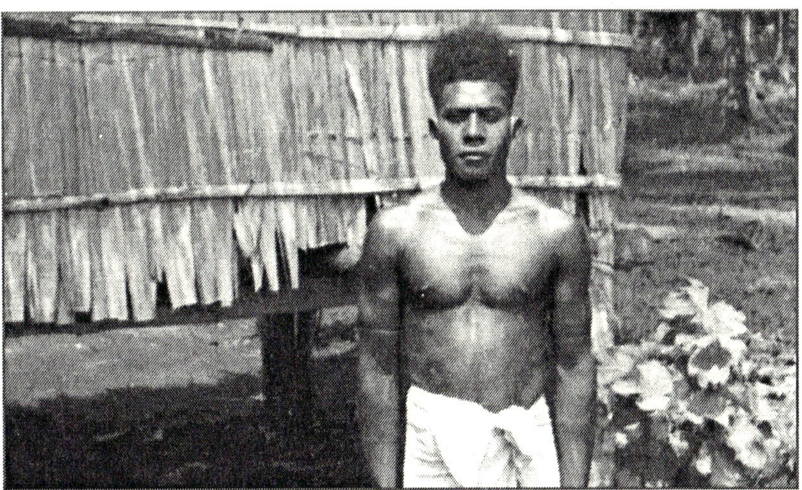

Both sexes pierce and stretch their earlobes and may put reed rings or shells on the loops formed. There is also a tradition of tattooing that has become less common.

I took pictures of the people, the villages and the landscape throughout the trip. Unfortunately, because of tropical conditions, humidity and the passage of fifty-some years since, many of the photos have not survived in the condition we would hope to see. I have included them, as they are, to help the reader better visualize the story.

Chapter 1

Fig. 1.1.
Ward Goodenough,
Daris Swindler and
Ann Chowning,
boarding a plane
in Philadelphia, as
photographed for
a local newspaper

It is February, 1954, and we are leaving Philadelphia for New Britain Island (Fig. 1.1). We will cross ten Time Zones as well as the International Dateline to get there with many stops between flights, a lengthy and tiring trip in the period before commercial jet planes.

Our first stop is familiar enough: Pittsburgh, Pennsylvania, where we spend just enough time for me to de-board, run to a phone booth, and call my parents who live about 80 miles south in Morgantown, West Virginia. Of course, they know about the trip. I am simply letting them know that we are on our way. I had already said goodbye to my wife in Philadelphia and my enthusiasm for the trip is tempered a little by the thought that it will be some time before we hear from each other again. But I am very excited nonetheless.

Next we are off to Chicago, Illinois for an hour-long stop, then on to Denver, Colorado. After a short stay we leave for San

Francisco, California, our last port-of-call in the U.S. I am awake all the while we fly over the beautiful Rocky Mountains. We are only a few thousand feet above them and have an excellent view of these magnificent mountains all covered with snow this time of year. After crossing the Rockies, I manage to fall asleep, and wake again only when the plane's engines begin changing tempo. There is a subsequent rumbling in the plane as the wing flaps are lowered down into the wind stream. We have started our descent into the airport in San Francisco. A layover of a few hours is scheduled before we board a Pan American Clipper for the flight to Hawaii, and when we finally do board the plane our flight is delayed for several more hours by engine trouble. We try not to think about this development.

By the time we finally take off we are ready for a cocktail and cigarette to help us relax and enjoy our rather long flight to Hawaii — about 12 hours in those days. I reach for my pack of Lucky Strikes and light up. In Honolulu we will spend several days visiting with anthropologists at the Bishop Museum, one of the world's finest Museums for studying the peoples and cultures of the Pacific. I hope we would also have some time to enjoy the beautiful beaches of Waikiki.

We stay four days, most of our time at the Bishop visiting and talking with anthropologists about our pending stay on New Britain. Dr. Goodenough has made several anthropological expeditions to the Pacific, but this is the first for Ann and me. We enjoy our visit to the Bishop Museum with all of its anthropological displays from the various Pacific islands, particularly the islands making up the Bismarck Archipelago, where New Britain is located. We are invited to a dinner party at the home of Dr. Kenneth Emory, an Archaeologist at the Bishop. It is a wonderful party with food consisting of various Hawaiian dishes, each with a different flavor and consistency. This is my first contact with poi (one of the principal foods of the Hawaiians) that is made from taro, and I find it too starchy and sticky for my taste. We are told many colorful stories that evening by anthropologists who had lived with native peoples on different Pacific Islands before and after World War II, stories that will likely prove helpful once we become residents of New Britain.

We leave Honolulu on a beautiful morning heading southwest over the Pacific Ocean on the Pan American Clipper *Bald Eagle*. Our destination is a coral atoll, the first stop on our long flight to Australia. The day we land the sky is clear, the sun hot, and the white coral sand even hotter. We spend about two hours while the plane is checked out and refueled, but there is nothing for us to see but sand and sea all around. We are glad to take off on the second leg of our flight across the Pacific.

This time we are headed for the airport in Suva, capital city of the Fiji Islands. On the way the *Bald Eagle* carries us across the International Date Line at 2220 Greenwich time. It is the 18th day of February, 1954. Before we land, the crew presents each of us with a certificate from the airlines on which was written:

That by so crossing this divider of days, the Today of mortals at once becomes Tomorrow and all is confusion.

There are about 250 islands of varying sizes ranging from large mountainous islands of volcanic origin to small coral atolls. Fiji is often referred to as the "crossroads of the South Pacific." Suva is on Vita Levu, the largest island in Fiji, and should be very interesting, but unfortunately the airport is some distance from Suva so we have no chance to tour the city. The stay in Fiji is pleasant, though, and we have several cooling drinks at a lovely bar at the airport and talk of our coming adventure.

The climate of Fiji is of a tropical oceanic type with southeast trade winds, and while the temperature was moderate, the high humidity was somewhat enervating. When it is time to leave, we board the plane with some fervor since we know it is only some 1,700 miles more to Sydney, Australia, where we will spend several days meeting more people, resting, and making the final arrangements for our departure to Melanesia.

The flight from Suva to Sydney is uneventful and time seems to pass slowly until the pilot announces that we are approaching the Great Barrier Reef off the northeast coast of Australia. The Great Barrier Reef is the largest structure ever built by living organisms, extending for about 1,250 miles off the northeastern coast of Australia. Many parts of the reef are dry or barely awash at low tide, and as we fly south along some of its length we have a clear view of the reef and the many fishing boats that daily ply the waters around it. It is a magnificent sight, one I am sure I will always remember.

It is summer in Australia, a marked contrast to the winter conditions we left behind in Philadelphia. The first night in Sydney is clear, and as I look out the hotel window I catch my first view of the Southern Cross, a constellation of stars I had read about many times but never seen. Seeing it made me realize just how far south we had come.

We plan to spend a week in the beautiful city of Sydney taking in the sights, visiting at the university, and of course, taking care of various chores. I will buy my dental supplies here, for example, for the dental impressions and casts I hope to make of the local people we will be living with in New Britain. During this week I will also fly to Melbourne, located at the head of Port Phillip Bay in south central Australia, to visit Dr. Roy Simmons, a hematologist at the Commonwealth Serum Laboratories. He is an internationally known investigator who had studied the blood groups of Pacific islanders for years. I had written him earlier about my plan to collect blood samples in New Britain, and he offered to supply me with vials containing the anticoagulant necessary to preserve samples until they can be flown to the laboratory in

Melbourne for analysis. I enjoy a pleasant two-day visit with Dr. Simmons that includes a day-long picnic trip to the Nandenong Mountains with his family. The next day I spend in his laboratory learning as much as I can about the problems of collecting and preserving blood samples in the field. We are to have a kerosene refrigerator that will at least keep the blood samples cool, but the anticoagulant will help immensely.

I return to Sydney and spend the next few days with Ann and Ward doing some last minute shopping and chores before our departure for New Guinea. One chore for me includes a visit to a Police Station. I plan to take fingerprints while in New Britain and hope that I might obtain the necessary materials there in Sydney. Once I explain about my studies, what I have in mind, the sergeant is quite helpful. He demonstrates how to make the prints and gives me a fingerprint pad and a container of ink. I offer to pay for the materials but he declines pleasantly, saying that this is the first time the Sydney Police Department has ever had such a request. I thank him and hurry back to meet Ann and Ward at the hotel for our farewell supper. We leave the next morning by plane for our next stop, Port Moresby, New Guinea, where, finally, we would be in Melanesia.

Port Moresby is on the southeastern coast of New Guinea facing the Coral Sea. When we land the weather is beautiful, hot, and even more humid than in Fiji. We remain in Port Moresby for more than a week at the Papua Hotel (Fig. 1.2), a two-storey structure built at the corner of two city streets.

Fig. 1.2. Papua Hotel in Port Moresby

Fig. 1.3. Policeman directing traffic

The almost statue-like figure in the middle of Figure 1.2 is a policeman standing on a box directing traffic, highlighted in this enlargement (Fig. 1.3).

Our rooms are on the second floor, which we anticipate will pick up the noticeable breeze after sunset. It is a nice, clean hotel with screens on the windows since there is no air conditioning. Unfortunately, the mosquito netting provided to keep out those flying insects not stopped by the screens also interferes with our breeze and it is difficult to sleep.

Again our routine is to spend some days making contacts with local scientists and officials and ensure we have everything we need before our flight to Rabaul, New Britain. We are excited and the days pass quickly.

At last it is time to leave! This will be the final flight, the last after so many since Philadelphia. Once in Rabaul we will rent a boat to take us and our supplies to the Hoskins Peninsula on the north coast of New Britain where we will live with the Lakalai people for the next several months. My stay there will be the shortest of the three, only about five months since I have to start back to graduate school in September.

At the eastern end of New Britain is Blanche Bay, which opens into the large, deep Simpson Harbor. Together they form Rabaul Harbor. Active and extinct volcanoes flank the north and east sides of the harbor; indeed, the harbor was a vast volcano itself that exploded about 1400 years ago creating one of the finest harbors in this part of the Pacific. Rabaul lies between the large harbor and the steep hills to the east and north. The town was built by the Germans before World War I and served as the capital of New Guinea from 1910 until the administration was moved to Lae, New Guinea in 1941.

We arrive in Rabaul after a rather short flight from Port Moresby and spend the next two weeks seeing people (we are getting used to this by now), buying more supplies, and arranging for a boat to take us to the Hoskins Peninsula and the Lakalai area. It is hot (90 F) and humid here, as we are only about four degrees south of the equator. By now, fortunately, we are beginning to adjust to the tropical weather. Many restaurants have outside eating facilities where, after "working" all day, we have a nice dinner and a few cold beers before retiring (Fig. 1.4).

Fig. 1.4. Café and bar in Rabaul

It takes several days, but we finally locate an appropriate vessel to take us to the Hoskins Peninsula: a sixty-five foot cargo boat that carries supplies and mail once a month to several plantations along the north coast and beyond that to Talasea, a small town on the northeast coast of the Willaumez Peninsula. The owner agrees to take us and our supplies on the next trip which is scheduled for March 10 and will deliver us across the bay from Talasea to the Hoskins Peninsula.

After almost a month of travel, tomorrow we begin the last leg of the trip to our final destination where we will finally meet the Lakalai people. I am restless with excitement. One more time I make certain all my data-collecting equipment is in order and collect my notebooks and pencils. I only close my eyes for a little while; my mind will not rest long enough for me to get to sleep.

Chapter 2

We left Rabaul this morning at 7:30 a.m. on the *Balus*, bound for Talasea. The captain, Steven Ephoff, is a local resident of Rabaul. The engineer is a middle-aged Chinese man who, we found out later, is also the chef. The crew of fourteen is made up of local natives (Fig. 2.1).

Fig. 2.1. Crew of the Balus.

The boat was loaded with supplies for the general store in Talasea and sundry items for plantations along the way as well as with our gear that included medical supplies, trade tobacco, bicycles, canned foods, and clothing. We were told by Captain Ephoff, that the boat would re-supply us "about every month or so."

The name of the boat, *Balus*, is the Melanesian Pidgin-English word for pigeon or airplane. Melanesian Pidgin-English is a true Melanesian language derived almost exclusively from Melanesian and English, and of course, WWII brought Japanese words into the vocabulary. The language has a synoptic vocabulary of more than 1,300 words and through them it is possible to translate some 6,000 English words.

The *Balus* is sixty-five feet long and twenty feet wide and has a wheelhouse, one passenger cabin, and a small well deck aft; the crew quarters were forward. According to Captain Ephoff she is able to cruise at four to six knots in good weather. We also found out that this was the captain's first trip as skipper, since his predecessor had been relieved of his duties and hospitalized in Rabaul. The previous captain, we were told, had been an Australian Air Ace in World War I. After the war, he came to the islands. He stayed and captained this vessel for many years before being overcome by effects of his war experience.

The passenger cabin is about seven feet by four feet with two bunks. There are cockroaches aplenty running around in the cabin, several quite large. As the weather was calm and the temperature warm today, we decided to sleep outside on the deck.

The *Balus* is steered by an interesting, if not amazing, arrangement of chains, gears, and wheels. Two chains run from the wheelhouse along both sides of the deck in a trough and connect aft to a perpendicular wooden shaft attached to the screw. It works fine until the boat has to turn sharply and then, because of slippage of the chains somewhere along their tortuous course, it is necessary to spin the wheel quickly several times in order to move a few degrees. This action results in a to-and-fro rocking motion of the vessel. In time, we became use to the rocking and all was well.

As we sailed out of Rabaul through Simpson Harbor we passed Mount Matupi, an active volcano near Mount Vulcan, a cone-shaped volcano several hundred feet high on the west side of Blanche Bay. Mount Vulcan is what remains of an eruption from the floor of the harbor in 1937, and this morning steam was coming from its sides, though the last full eruption of these volcanoes had been in 1941 (Fig. 2.2).

Fig. 2.2. Mount Vulcan in Blanche Bay

We left Blanche Bay and sailed along the northwest coast of New Britain passing the South Mother, another active volcano, before we reached the North Baining coast, named for the Baining people who inhabit the northern end of New Britain. At dusk, the sea was beginning to get a little rough. The captain took a bearing on the lighthouse at Cape Lambert Point, and headed southwest into the open sea in the direction of Talasea. We had dinner and by then the sea was becoming quite a bit rougher and waves were mounting higher and higher up the sides of the boat. It was then that Ward and I decided to go to the wheelhouse and find out what was happening. Ann was becoming seasick and retired. The captain told us that we were in for a good blow.

We remained in the wheelhouse with him until about 9 p.m. when Ward, not feeling too well either, went to bed. The captain told me I could use his bunk since he was going to spend the night in the wheelhouse. I went to bed about 10 p.m., but in a short time, the boat was rocking so violently I could not sleep so I re-joined the captain in the wheelhouse.

By midnight the storm was so strong and the waves so high that the boat bobbed around like a cork. I was in the U. S. Navy in World War II, in the North and South Atlantic, and had been through several storms as well as a hurricane so I appreciate the power of storms at sea. Even so, I was becoming alarmed. This was not a WWII tanker, but a much smaller boat and the storm was becoming quite bad. The captain had to abandon his course and concentrate on heading into the waves rather than allowing them to strike our beam.

Just as the captain was telling me that this was one of the worst storms he had seen the engines stopped and we were adrift, at the mercy of the sea. The engineer worked like a mad man. There was a lot of shouting and clanging of wrenches against metal. He seemed to know what he was doing, however, and finally the engines started up again. We continued turning into the waves for what seemed a long time, and even though the storm eventually subsided, the waves remained rather high throughout the rest of the night. Our boat proceeded rather slowly for another few hours and eventually resumed course. Needless to say, there was little sleep aboard the *Balus* that night.

March 11

No one wanted breakfast this morning and none was prepared. By noon, however, things were looking a little better and we had sandwiches. We had lost so much headway, though, that by evening we were still only a little off the Northeast coast of the Willaumez Peninsula and some distance from Talasea. The captain decided to find a cove and anchor for the night because of the many uncharted coral reefs in the area. As we passed between two small islands one of the crew caught a large mackerel weighing about 25 pounds. We anchored in the lea of the southeast coast of the Willaumez Peninsula and prepared to eat mackerel and spend the night.

The engineer, as I mentioned, also doubles as the cook. He cut the head and tail off the mackerel and gave these to the crew who prepared their own meal. The rest of the fish he cut into thick steaks that he said he would prepare "Chinese style" with rice, cabbage, a few spices, and some unknown (to us) condiments. This was our first real meal in 24 hours and it was delicious. We thanked the cook and sat on the deck enjoying a refreshing beer after dinner. Everyone would sleep well tonight.

March 12

We pulled up anchor at 6 am and set out for Talasea, a small town supported mainly by several plantations in the area. As the captain had said, there are many coral reefs guarding Talasea harbor but we managed to snake through these beautiful but treacherous outliers and dock about 9:30. A group of local people had gathered at the pier to welcome us to Talasea, including the Assistant District Officer, Mr. and Mrs. Michael Foley, Medical Officer A. V. "Bill" Bell, and Mr. and Mrs. Gorden, the local school teachers. Much to our surprise they also had received mail for us, our first since we had left the States. We spent an enjoyable morning talking with Mr. Foley, after which, we lunched with the Gordens. Later we visited Bill Bell at the local hospital. Bill had come here shortly after World War II and dedicated himself to a life of medical work in New Britain. He told us he had never had a boring day since coming here, and it was obvious that he was well adjusted to life in the islands. I asked him if I might borrow a microscope from him with which to examine blood smears for sickle-cell anemia (a condition associated with malaria). I also needed a solution of 10% formaldehyde and was happy to find that he could provide both.

This evening we dined with Mr. and Mrs. Foley and afterward spent several hours discussing the local native situation over a few beers. It is obvious that Mr. Foley was very concerned about native matters and quite aware of the many problems facing the native population. It was soon made apparent to us that two major issues facing the region were the increase in the native population size and land tenure, the latter including the need to get local populations interested in producing more in their gardens than they required for their own sustenance. As is often said, a solution for one group does not always apply to another group and our conversation with Mr. Foley made it obvious that this saying applied to the people of New Britain. Indeed, no solutions were proffered tonight, but we better understood some of the many problems that the Lakalai were facing. It was a most interesting evening. Ward and I returned to the boat and Ann stayed on shore for the night with the Gorden's.

March 13

We sailed this morning at 6:15 from Talasea for the Hoskins Peninsula, the home of the Lakalai. We are excited, for at long last we were going to meet the people we have come so far to live with and study. I am especially anxious since my studies as a physical anthropologist would involve physical contact with the people, something they may or may not permit. I plan to take measurements of their bodies, prick their fingers for a few drops of blood, make finger print impressions, take samples of their hair and, hopefully make dental impressions of their teeth. I have vials for preserving blood, a head spanner for measuring head height, calipers for measuring facial and head dimensions, an anthropometer for measuring body dimensions, and several pounds of Geltrate (impression material) and Albastone (casting material). I will have to wait to see if I am accepted and allowed to use any or all of this material. I was probably more anxious this morning than my colleagues who, as cultural anthropologists, were to investigate various aspects of the culture e.g., social structure, kinship relationships, language and religions, none of which required physical contact with the people.

Before reaching our destination, however, the *Balus* had to deliver supplies to two plantations. As we approached the first plantation the boat sailed into a narrow cove and almost immediately there was a bump as the boat ran up onto a fringing coral reef and stopped. The captain reversed the engines and , thankfully, we were able to sail back off the reef without damage to the boat. We proceeded in and offloaded supplies. Only one more stop at another plantation before we proceeded to the Lakalai coast.

Finally we arrived at The Hoskins Peninsula (Fig. 2.3).

We disembarked and were met on shore by a group of Lakalai men who assisted us in getting our supplies off the boat and storing them while it was decided in which village we would live.

One of the men meeting us was Boas, the *luluai* (government appointed headman of a village) of Galilo, the largest village on the Lakalai coast. As Ward speaks and understands Melane-sian Pidgin-English well he had a good chat with him about our

plans to stay in his village. Boas seemed pleased with this plan, and although he appeared to be a little surprised at the proposed length of our stay, the matter was settled. We made an appointment to visit him again the next day and discuss our plans in more detail.

A little later, Miss Cynthia Smith, the nursing sister at the Malalia Mission in this region, came by to say hello and welcome us to the island. She suggested we stay in one of the Mission houses for the night. We talked for several hours with Sister Smith who was obviously quite knowledgeable concerning the Lakalai, and though reluctant to stop, we were obviously exhausted and Miss Smith politely suggested we retire. Ward and I were to sleep in one of the houses and Ann would stay at Miss Smith's house. Since tomorrow is Sunday, Miss Smith invited us to attend church with her.

The church is an interesting building about thirty feet long by twenty feet wide, constructed of thatch (branches and leaves of local plants e. g., the sago palm) and built over a concrete floor. The construction is similar in style to the *singsing* house we had seen in Rabaul. (A *singsing* is a song, party, or dance usually associated with a ritual or ceremony of some sort.) Three vertical poles in line with each other extend above the roof of the church, one in the front, another in the middle, a third at the rear of the church. The service this morning was given by a native minister who spoke only Lakalai. We couldn't understand a word, of course, but hummed along during the singing parts. The men and boys sat on chairs while the women with their babies as well as young girls sat on the floor. Miss Smith and Ann sat with us in the chairs.

After church we returned with Miss Smith to her house and were shortly joined by four men from the village of Rapuri that included Lima, the *luluai* of that village and spokesman for the group. After talking for a short time, Lima invited us to live in Rapuri during our stay with the Lakalai. Ward explained to him that we were going to visit all of the Lakalai villages during our stay but would probably stay in Galilo, the largest of them. This seemed to satisfy Lima and the other men, and after a little more *toktok* (talk) they left.

I should mention that there are two Christian religions represented among the Lakalai — Methodists and Catholics — with only one denomination present in a village. The first permanent missionaries settled in the Lakalai region in 1922, and subsequently the Lakalai joined either the Roman Catholic or the Wesleyan Methodist Church. Lima represented a Catholic village. Because Boas's village was Methodist, we are a little concerned that our choice of villages might have undesirable ramifications between the two. We hope for the best.

After they left Miss Smith informed us that she believed Lima was one of the leaders of the local Cargo Cult which had developed in Melanesia between the two World Wars. The cult had all but collapsed during the Japanese occupation only to reappear at the end of the war. The common theme in a Cargo Cult is that the goods and machines of the white man are actually manufactured by the ancestors of the islanders in the land of the dead. Later, they believe, the goods are brought to the native populations by airplane or by ship. This belief came from observing the supplying of material by air and sea. The Europeans, the cult members assert, intercept these materials which become the source of their power over the natives. The cult maintains that one day the ancestors will return to the islands to make things right, but in the meantime, in order to receive the cargo, the island people must have faith in the new cult and wait for goods to arrive, even if it means neglecting their own means of sustenance. In extreme examples of this cult, though this aspect is seldom seen anymore, cult members destroy all their own possessions and simply wait for new goods to be delivered. The final consequence of the Cargo Cult is the return of the dead and the arrival of great cargos to the islanders followed by the overthrow of the white man's rule.

Miss Smith said Lima is also considered controversial by the Australian government. We had been aware of the Cargo Cult in Melanesia and some of the problems it poses for the government and natives as well, and as anthropologists were very interested in the fascinating topic of the Cargo Cult and its underpinnings in Melanesia though we, ourselves, wished to avoid controversy. Lima, we think, might prove to be a valuable source of information.

This morning we rode our bicycles to Galilo, a distance of about five miles over a trail that passed along the beach and meandered occasionally into and back out of the jungle just a few yards inland. It was our first experience at cycling on a tropical island and it was quite an experience! We had to stop to rest several times. The beach portion was sand and the jungle was riddled with tree limbs, roots, coconuts, and bumps of all sizes. For three novices, the ride proved somewhat difficult. Eventually we reached Galilo, and Boas met us at the outskirts of the town. He was pleased to see us and immediately took us to the site of the *haus kiap* (house built for a Patrol Officer) that the village men were building for us. The project is in its final construction phase and they were putting the sides of the house together. The sides consist of long bamboo poles, each about six feet long and wrapped in sago leaves, each leaf placed so that it overlaps the one above. The leaves are then sewn together by long threads of sago and each completed section placed on the side of the house, not unlike modular house construction we see today in the States.

Fig. 2.4. Our house in Lakalai. The small building to the left of the house is the "outhouse" built especially for us.

Boas told us that this was our house for as long as we stayed in the Lakalai region. It is a spacious dwelling compared to the other houses in the village, with two large rooms, a kitchen and a porch (Fig. 2.4). Situated on a knoll overlooking the beach and the sea, it is a comfortable dwelling. After looking the construction over and telling Boas we would be happy to stay in the house, we sat on the porch and had our first taste of *kulau*, (the milk of a green coconut), a cool and quite refreshing drink.

After Boas left, we walked to Rapuri to visit with Lima and some of the elders in his village. We had a long talk with Lima who said he was disappointed we had not let him know of our visit since he would have prepared a feast for us. He also told us that he was also having a house constructed for us and there were many advantages for us if we stayed in his village as he was the head *luluai* of the Lakalai region and that Boas only had followers in Galilo and three other Methodist villages. We suddenly realized that we were caught in the middle of a politico-religious friction among the Lakalai, and that two important *luluais* were vying to get us to stay in their villages during our time in Lakalai.

We explained to Lima that we are neutral in our position and interested in all the Lakalai, not in any one village or region. He seems to understand our position but it is clear that both Lima and Boas are powerful leaders and we may have more problems of this nature later. It is also very clear, however, that both men want to be friendly and are willing to help us in any way they can. After a good bit of conversation we left Lima and walked back to Galilo. During a short visit with Boas, Ward confirmed that we were staying in Galilo. Boas was very pleased and said he would have some of the men help us collect our gear and bring it to Galilo. We left by bicycle for Malalia in a pouring rain (we were still in the rainy season that lasts from December through March). It was getting dark, but we arrived in time to have dinner with Miss Smith.

We are retiring early, looking forward to what the next day might bring.

March 16

Unfortunately, today it also rained and the move to our new house in Galilo was postponed. Ann and I remained in Malalai studying Pidgin, while Ward cycled to Valoka to visit Father Berger and the Sisters at the Catholic Mission.

March 17

Today was beautiful. The sun was out and the jungle dried out quickly since the soil was mainly volcanic and drained rapidly. I went with Miss Smith to her clinic where she held "out-patient" hours every morning. Most of the patients were women and children since the men only come to the clinic when they were very ill or suffering from wounds. Cuts are the most common complaints at the clinic since tropical ulcers can occur from even the smallest scratch. Sulfur powder is used on all open wounds. It is remarkable to me how the children remain so calm during treatment, no matter how severe the injury and in spite of their obvious pain. This is true even when they have to have stitches, or when iodine is used instead of sulfur powder.

Miss Smith also told me that the local women were up within a few hours of giving birth and in a few days they were back working again in the gardens. After a most interesting morning spent with Miss Smith in the clinic, we had lunch together and I returned to the house. We had hoped to move to Galilo today, but the village was having a feast which meant that the men would not be working. We will stay another night at the mission.

March 18

Today it rained all day but in the early afternoon three outrigger canoes came from Galilo for our gear. We loaded everything on board and they took off, Ward cycling to Galilo to see that everything arrived safely. After his return we spent the rest of the day studying Pidgin and watching it rain.

March 19

It rained all day. Nevertheless, we cycled to Galilo and arrived soaking wet to begin arranging the house that would be our home for the next several months. Not long after we arrived Boas and several men came to help us unpack. We had a great time unpacking, showing them what we'd brought, pointing out where things should go. It was fun and by late afternoon we were pretty well ensconced in our new home. Boas and his helpers left laughing and singing. It was a day enjoyed by all. And it was still raining when we retired after dinner.

I notice that a large spider has taken up residence on the outside of my mosquito netting. I hope his only designs are on the mosquitoes.

March 20

Our house is built on a slight bluff above the beach a short distance outside Galilo and, from the porch, has a beautiful view of the Pacific Ocean. We soon learned to appreciate the soothing sounds of the waves splashing against the beach since they helped us fall asleep. Ward and I live in the front room and Ann has the rear room. The outhouse is a short walk behind the house. Two men from the village do our cooking: Uoso the "chief" cook (Fig. 2.5), and Sege, somewhat younger, his able assistant (Fig. 2.6).

Both these men proved to be excellent cooks and became close
friends and great companions as well as reliable informants of
the "doings" in Galilo.

Fig. 2.5. Uoso

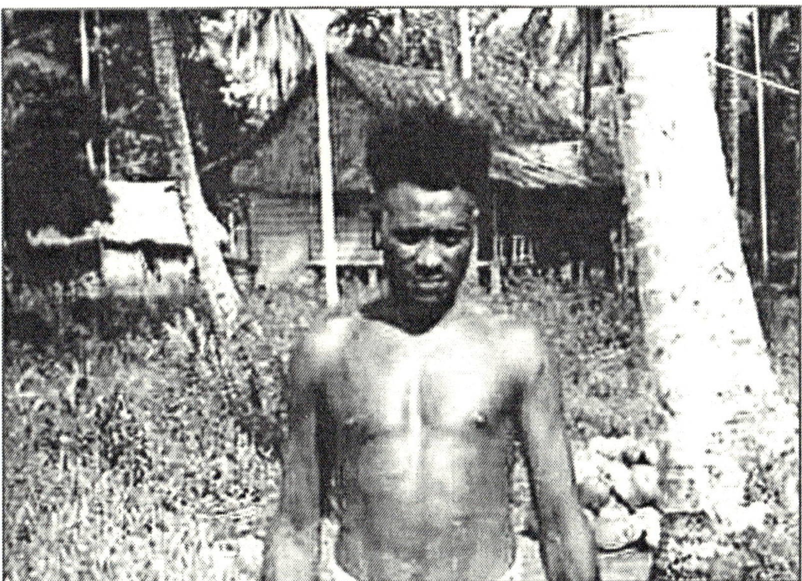

Fig. 2.6. Sege

Today I unpacked my anthropometer and calipers on the porch and demonstrated their use on several male "volunteers" as people gathered around to see what was going on. There were gasps and several *oh's* from the onlookers as I took stature and head length measurements. It was obvious to them there was no pain involved in the process, and everyone seemed to enjoy watching me work. After this show, Ward, who had also been watching, suggested I prick his finger for a drop of blood to show everyone that there was no pain in the blood-collecting procedure either. It is perhaps too soon to draw blood from the people, but we thought this early demonstration might help lay the groundwork for this later stage of data collection.

Boas was one of the attentive observers this afternoon and I asked him if he would like to see the blood sample under the microscope I had borrowed from the clinic in Talasea. He said yes and when he peered into the microscope he at first looked somewhat confused, then smiled from ear to ear as I explained that the blood of all people looks alike. Several other men came forward to have a look and one or two offered their fingers. These little demonstrations proved an excellent opportunity for me to show the people what I wanted to do and gain their confidence. I was sure that word of this afternoon's activities would be passed among the villages.

After Boas and the other locals left, we walked to Rapuri in the pouring rain to visit Lima. We had brought *tabab* (stick tobacco) from Rabaul to be used in Lakalai, and while Ward talked of what we wanted to do while in their country we offered a stick to Lima and the several men who gathered to listen. Lima invited us to stay for dinner and, of course, we did. We had taro, the main crop of the Lakalai. This tropical Asiatic plant (*Colocasia esculenta*) is used in Polynesia to make poi. The tuber was cooked in coconut milk and was more palatable than the poi we had in Hawaii. There was so much of it, yet we kept eating, feeling we had to put away what the women kept bringing to us. After a while, my stomach could not hold any more, and about then, thank goodness, the women stopped bringing us taro. We had a cup of coffee and continued our conversation with Lima. He talked mostly about his company that sells copra (dried coconut meat that is the source of coconut oil) to Mr. Maynard at Matavulu, who in turn, buys tin that is then used in house construction instead of thatch. This company, it seems, might be in conflict with the co-operative stores set up by the government. This should prove interesting ground for future study. By the time dinner was over it was getting dark. We said our good-bys and headed back to Galilo in the pouring rain.. Fortunately, we were met about halfway by Sege, and we followed his flashlight back to Galilo.

March 21

We were up early and went to church. The church in Galilo was similar to the one we attended earlier with Miss Smith. In the afternoon, Boas came by and we walked through Galilo for the first time. Galilo is the largest village in Lakalai. It is kept clean by the women and children who pull the grass and sweep the ground each day with thatch brooms. There are men's houses in addition to family houses and cookhouses in the villages. The men's houses are on six-foot poles with benches underneath to sit on. We are told that in the "old days" the men spent most of their nights there, even when married. The cookhouses are built on the ground while family houses are set on two-to-three-foot poles (Fig. 2.7).

Fig. 2.7. Three boys standing in front of a raised family house

Pigs, dogs, and chickens have the run of the villages and are everywhere, grunting, barking, and clucking. Boas told us that in the "old days" men from Galilo hunted cooperatively with several other villages for bush-fowl eggs in the somewhat distant thermal areas where they are found (Map 3). Only men participated since more distant villages might be hunting eggs in the same territory, resulting in fights in which, sometimes, men were killed. He recalled as a young boy having fights with other villagers. These days, he said, there are no more fights, and women and children are safe to go collecting eggs.

After a long day we came back to our house, had dinner, and studied Pidgin before retiring.

March 22

This was a beautiful day. Ward and I got up early and cycled to Malatlia to meet Miss Smith and to see if we had any mail. There was none, though, and we returned empty-handed to Galilo.

During the afternoon we studied Lakalai for a couple of hours. Sege came by and we talked about what I was to do while in Lakalai. I explained that my work would entail measuring many people in all villages, and I suggested that perhaps he would like to be the first person I measured. I explained to him the importance of my studies of his people and that, if he permitted me to measure him, others in the village might let me measure them also. He said he understood and agreed. I got out my calipers and set about measuring my first Lakalai. Sege was an excellent subject and as people gathered to see what was going on he helped by telling them they should let me measure them. I would tell Sege in Pidgin what I wanted them to do and Sege would translate it into Lakalai. I would only be measuring males at this time.

March 23

We arose to another lovely day and since it is nearing the end of the rainy season, we can expect better weather. We spent the morning studying vocabulary and practicing Lakalai with a group of women and children that came by every morning after breakfast to sweep the ground around our house. Their brooms are made of various flexible plants and sticks tied into bundles with grass rope. Each broom is about two feet long and does an excellent job.

It is very pleasant having people come by, especially the children who are always eager to sit and talk to us. They also want to teach us their language and have great fun trying to have us pronounce the words correctly. Ann, I must say, is always way ahead of me when it comes to learning the local language. This is very important for her since women are not permitted to speak Pidgin and one of Ann's studies is to work with the women and study their contributions to the Lakalai society.

After lunch I had my first fishing trip with Sege and several men from the village. On this trip we used a line wrapped around a bottle with a wire leader and hook attached to the free end of the line. I asked Sege about using a fishing pole and he said some men used a pole but he likes a hand line better. He told me that *pis spia* (fishing spears) are often used and one day we would go fishing with spears. Today we rowed out to a coral reef in an outrigger canoe; my first trip in an outrigger. The canoe is about fourteen feet long with five wooden poles, each about eight feet long, projecting from the canoe to the parallel outrigger. Three cross poles are tied to the outrigger and everything is lashed together with bamboo. The three-foot long paddles were broad at the water end and pointed. The outrigger is so constructed that either end can be the bow. Sege and I paddled and I was aware how easily the outrigger glided through the water with little effort from the paddlers. I was told that when traveling near the shore they often use a long pole eight to ten feet to push the outrigger along. Today the weather was murky and not clear enough for fishing, so we returned to the beach. It was an exciting experience, one I thoroughly enjoyed, and I look forward to another fishing trip in the near future.

I retired soon after dinner and as I prepared for bed I noticed my friendly spider still in position and watching over me.

March 24

The weather was clear and warm this morning and we had a quick swim before breakfast. After breakfast, several men came by to be measured. I gave each person one stick of trade tobacco, the amount we had agreed to pay informants, and continued this practice throughout the project. Shortly before noon, a group of women came by with eggs from the bush-fowl. The eggs are hatched in thermal regions (Map 3) at various depths.

The Lakalai women lower themselves into these burrows to collect the eggs. The birds are known as *faul bolong bus* (bush-fowl) and their eggs are an important dietary supplement during the egg-lying season. The eggs are about three to four inches long, light brown and very tasty.

The women also had one cassowary egg with them. We bought all of the eggs and had the cassowary egg for lunch. Each egg is large, green and about six inches long—enough for the three of us. It tastes like a chicken's egg and we told Uoso that we would like to have them again. He said he would try to get us more, but it is difficult since the cassowary is such a dangerous bird.

I measured several more men during the afternoon and by now had no trouble finding recruits. I wanted to wait a while longer, however, before requesting blood samples or invading their oral cavities for tooth impressions.

For dinner we had a delicious fish cooked in coconut milk with mashed taros after which we demonstrated the tape recorder to Sege and Uoso. Uoso recorded a short conversation with Sege and, when they heard their voices, they began to laugh and say, "oh! oh!" and shake their heads. They were fascinated and called it "the box that steals your voice." We suggested that we would record the next *singsing* in the village but Uoso told us that it would be awhile before there was another one. A pig, he said, belonging to a *luluai* had "gone bush" (gone crazy) and been killed and even though this happened in another village, the people of Galilo are in mourning. He did not know how long the mourning would last.

March 25

Today we left after breakfast by bicycle to visit Father Berger and the Sisters of the Valoka Mission. It is some distance from Galilo, and by noon we had gotten only as far as the Government Station near Cape Hoskins where Patrol Officer Ernest Sharp and his wife, Mavis, live. They invited us to stay for lunch, and of course, we accepted. For lunch Mrs. Sharp served Australian Bully Beef (not unlike American Spam), large sweet potatoes, and a green salad; it was delicious. During lunch Ernest said he is planning a trip of five weeks into the interior mountains of the island. He invited me to accompany him. The trip will be mainly to take a census of the people living in the villages but at the same time he would be searching for a man, a reported sorcerer from one of the villages, who had apparently killed another man. The trip is scheduled for shortly after Easter. Ernest has been in the mountains before and stressed it would be a rugged and demanding trip. I had read about the geography of New Britain earlier and know that the mountains run from one end of the island to the other and that this insular divide stands from about 5,000 to over 6,000 feet in elevation with few exceptions. Several volcanoes are strung out along the northern perimeter of the island and many of their slopes merge with the interior mountains and extend to the sea. The highest peak in the island is a 7,500 foot high active volcano on the Northwest coast called The Father. The mountains are geologically young, preserving sharp ridges, precipitous slopes, and deep V-shaped valleys. He said he found that the mountain people were usually shorter and somewhat hairier than coastal people, and some groups bind their babies' heads so that they grow in an elongate shape. Further, the mountain people are shy compared to coastal people, he said, since they have not had as many contacts with foreign people. He doubted if any of the villagers would let me measure them, especially since we would stay in a village only a day or so taking census data. All of this is exciting to me as a "budding" anthropologist, and I said I would be happy to join him.

We left the Sharps and continued cycling to Father Heinrich Berger's Mission. A short distance from the Mission we were met by a number of *luluais* from several Catholic villages who had been organized by Lima to meet us. They had been waiting all morning so we stopped and shook hands with them while Ward explained what we intended to do during our stay among the Lakalai. They seemed satisfied and said they would cooperate with us and tell the people in their villages why we were here. This meeting left little doubt in our minds about the influence Lima had among the Catholic villagers.

Father Berger is a pleasant man, perhaps in his early 50's, rather tall, slender, and soft-spoken. He told us he had not been feeling well for some time, which is probably related to his malaria. During our conversation, he said Lima was most interested in the welfare of the Lakalai and always looking for ways to improve their lives. He also told us that, in his opinion, Boas is interested only in his own personal gain and not concerned with the welfare of the Lakalai. We thought it better to listen rather than enter too deeply into this conversation. Later, we met the Sisters of the Mission who were quite charming and very talkative. Their German order does not permit them to return home, so they spend their lives on the island, teaching and working with the people, running the hospital, and working with the Father.

At about 5 o'clock we said good-by to Father Berger and the Sisters and started cycling the ten miles back to Galilo. By now we were seasoned "jungle cyclists." We talked along the way about our conversation with Father Berger, and before too long, we were home. We had time for a short swim before dinner, after which we sat up discussing the events of the day and retired about 11 o'clock.

We were up early. After breakfast we had a visit from a local young man who had gone to Talasea a year ago and become a medical assistant. He was home for six weeks and, when hearing we were here, wanted to meet us. He was very interested in what I was doing and wanted me to measure him. Sege said he would record the measurements for me, which he did very well. The young man said he thought his people would let me measure them and that he would talk to them for me.

Boas came by later in the afternoon and said he would have five men tomorrow morning for me to measure and five more in the afternoon. My sample was increasing daily and I would soon see about taking blood samples and making dental impressions.

About 4 pm a young boy came by the house and asked me to go with him and the medical assistant to see a sick man in Vavua, a village about one mile from Galilo. The man has been sick since last December and recently has begun to have terrible stomach pains and difficulty in urinating and defecating. He was lying in his house when we arrived and did not look very well. When I touched his swollen stomach he said there was much pain. We filled a syringe with warm water and gave him an enema. Within minutes he expelled large amounts of fluid and some fecal material. We had brought erythromycin tablets with us, and gave him two. The medical assistant said he would spend the night with him so I left some additional tablets with him. The medical assistant and I both asked the man if he would go to Talasea for treatment but he said no, he wanted to die in his village with his family.

As I walked back along the beach to Galilo that evening, all that had happened today was swimming about in my head; thoughts about our earth and its many people, its many religions and cultures, and then I realized: this is why I became an anthropologist. This man was dying, he knew it, and would not leave his village or family. I believe I would have done the same thing under similar circumstances.

When I got back to Galilo, Ward and I went swimming and relaxed in the warm salt water of the Pacific Ocean. Coming out of the water we met Father Wagner walking down the beach. We had heard of Father Wagner but had not yet met him. He lives in the Central Highlands behind the Lakalai country and only comes to the coast on rare occasions. He is quite striking, about five feet five, slight of build, with a long, flowing beard that hangs almost to his stomach. He stayed for dinner after which we spent the evening talking till midnight. He doesn't trust Lima, he told us, and believes he is behind the Cargo Cult "outbreak" among the Lakalai. He said he prefers the mountain people to the coastal people and that is why he rarely comes to the coast. I found him a little strange, but likeable. His visit reminds us of the complicated alliances among island people.

March 27

Another day measuring, taking fingerprints and making observations of body build, skin color, and hair patterns. I worked all day and by 5: 30 had measured 20 men so I decided to go swimming with Ward and Ann, who had spent the day collecting genealogical data and Lakalai vocabulary. Father Wagner had gone to Voloka that morning and returned that evening with mail for all of us. It was nice to hear from home that everything was OK. As usual, that night my spider was waiting for me on the mosquito netting, the last thing I will see before falling asleep.

Father Wagner conducted Sunday Service today. The church was full, with women sitting on one side of the church, men on the other. There were four altar boys dressed in purple *laplaps*. Father Wagner stood out in his colorful vestments. The service was conducted in four languages: Latin, Lakalai, Pidgin, and English. There was much singing by the congregation in addition to the small boy-choir which sang several hymns. The service lasted about two hours. Afterward, we went back to our village and saw Father Wagner off as he left for the mountains. That afternoon we had visits from both Boas and Lima. Lima came by not long after we had returned from church, and we had just begun to talk when Boas came to the house. This is the first time that we had ever seen the two men together. They acted in a quite friendly manner, but it was apparent that neither was at ease with the other. Lima stayed only a short while longer, saying he had work to do at home. Boas stayed most of the afternoon.

Later, when several other men came to the house to talk, we showed them a map of the world, pointing out America and New Britain. They asked how far apart the two were, and we told them it was a very long way. It was obvious that they do not really understand how far, but it was an interesting afternoon and we all enjoyed ourselves.

Still later, a mother appeared with a child who had been stung by a jellyfish a day or so ago while swimming. His foot was very swollen and hot to the touch. I washed out the wound with warm water and pressed the ankle, trying to relieve the swelling and pain. A great deal of pus came from the wound. I applied a hot bandage, which seemed to relieve him, and gave his mother a few erythromycin tablets to give him.

I spent the day measuring several men in Galilo and noticed for the first time that some of the women were coming by to watch. By late afternoon I had measured thirty men from Galilo and was quite pleased. Soon, however, I would like to go to other villages for I want to include people from all Lakalai villages in my study.

Shortly after I measured the last man for the day, Boas came running up to me and said, very excitedly, that his young son had been attacked by a wild pig as he was walking back from the garden. Ann, Ward and I went quickly to his house. The medical assistant was waiting. The boy, about fourteen years old, had been seriously mauled by the pig. His right arm muscles had been ripped through, back to front, almost down to the humerus. There was a lot of blood. Below the right elbow, several muscles were exposed on both sides of the forearm. The left arm was OK except the hand had been badly chewed and we could see several punctures between the thumb and second finger. A split between the second and third fingers exposed the bones. Probably he had held up his arm to protect himself when the pig knocked him to the ground. Four wounds had been made on the chest, probably by the tusks. Another deep wound below the right knee had just missed severing the patella tendon.

Unfortunately, we had no sedatives except aspirin. We washed all the wounds with antiseptic water and set about suturing. Ann held up a lantern and a flashlight through the whole procedure so we could see what we were doing. Ward sped off by bicycle to Valaka, some ten miles away, to get medicine from the sisters.

We poured sulfur powder into each wound and carefully sutured it closed, working for about three and a half hours before we finished. The boy remained conscious the entire time, his mother beside him holding his right hand. He cried only a few times during this ordeal, generally remaining quiet.

Remarkably, the pig had missed the major arteries in his arm or the bleeding would have been worse. The punctures to the chest had also missed the intercostal arteries, though several veins in the chest were punctured and these bleed slowly. The wound to the arm was so wide, however, that we could not suture the skin together but had to wrap a large bandage around the arm to hold muscles and skin together. We bandaged the other wounds then took his pulse to see how he was doing. It was 78. Our pulses were probably much higher. Ward returned from the mission and we gave the boy a few pills. We were all tired, so we left, telling Boas and his wife that we would check on the boy again in the morning.

Just before coming on this expedition I had taken the first semester in the Medical School at the University of Pennsylvania that included human anatomy, so I was somewhat familiar with the anatomy of the human body. Now I was very grateful I had done it. It was still my first experience at suturing and applying bandages but the medical assistant had such knowledge and we managed it together.

We would hardly sleep that night for worrying about Boas's son.

March 30

I was up at 6: 30, had breakfast and went to see Boas's son who appeared to be doing well. As we changed his bandages he smiled and said he was feeling better. If no infections set in, he would probably be OK. Bill Bell had been contacted and sent word that he would come tomorrow and take him to Talasea.

I returned to Galilo and measured several men, something relatively uneventful compared to yesterday's activities.

This last day of March I spent measuring several Galilo men, bringing my sample to 41. During the day several people came by our house to have sores cleaned and bandaged. There are always many children about our house and I learn many Lakalai words from them since they are always eager to teach me.

Late in the afternoon, Boas came by with men from the East Lakalai, which interested Ward very much since he wanted to collect vocabulary from them in order to compare the two dialects. I was also interested since I thought they might be physically different from the West Lakalai. There were no major physical differences.

Chapter 3

April 1

Shortly after breakfast this morning I began to have stomach pains. I thought it must be an April Fools joke, but it wasn't. I went back to bed and slept till 10:30. After getting up I felt a little better but didn't feel like working, so I decided to stay home. Uoso and Sege were there and we talked most of the morning. I learned from them that the large, round baskets used for carrying articles to and from the gardens are woven by men only, while the small baskets carried by the women are woven by both sexes. The separation of sexes in many activities is common in Melanesia as in many parts of the world.

After lunch Sege and I walked to his house where I met his parents. Next we went to the family garden where he pointed out and named the various plants growing in the garden, e.g., taro, *bin* (green beans), *kaukau,* (sweet potatoes), and yams. It should be noted that the sweet potato is not indigenous to the South Pacific, but is a native of the Andes of South America. There would seem little doubt that either accidentally or purposefully the sweet potato was carried to Oceania by people from the New World. The timing of its initial spread into the Pacific, however, remains a mystery.

Sege showed me several other plants that were quite aromatic and explained that they were used by the men to make "love potions." The men make the potion and then rub it in their hair and on their chest. When a man wearing the potion walks by a woman he desires, when she smells the potion it is believed that she will want to follow him into the bush. I suspect that much of this depends on the woman's innate desire. Sege also showed me a plant used to poison fish. Material from the plant is taken to a reef and set afloat, and as it passes down along the reef it poisons the fish which then float to the top of the water and are collected. The poison is weak and the fish are edible.

All in all, I had a pleasant day even though it started out poorly. Some men of the village brought us several fish, one of which we had for dinner. It was delicious.

April 2

I spent the morning measuring and finger printing. My sample is now 50, and in one or two more days, I should finish my Galilo collecting and move on to other villages.

During the afternoon Ann and I saw ten or so people with various skin infections and gave them salves that they could apply themselves. Boas's son came by today and we applied clean bandages and sulfur powder. His wounds are getting better and there have been no infections though I didn't have a chance just then to check all the injuries.

We went swimming before dinner—what a refreshing way to end a day! After dinner, as usual, we spent the evening comparing notes and discussing the day's activities.

April 3

Today it rained, but I measured four men before lunch. People came to our house throughout the day for stick tobacco and to talk to Ward.

Later in the afternoon, I talked with Sege about the "love potion" he'd told me about which he had said was only used by the younger men in the village. I had some shaving lotion left and passed the bottle to him so he could smell it. He said he thought it would be very good to encourage the young girls to the bush and asked me if he could have some to put in his hair and on his chest. He patted some on himself and left exceedingly happy. I never found out if my shaving lotion was as strong as the local plants.

Over the rest of the day I measured several men who came by the house and volunteered. Soon I would start going to other villages where I hoped they would be as receptive to my work as the men in Galilo.

We had fish for dinner and talked for a while. When I went to bed my spider was waiting for me.

April 4

After breakfast the medical assistant came by for more sulfur powder to take to Boas's son. I went along with him to see how the boy was doing. We cleaned his wounds, powdered and re-bandaged them. The wounds were healing, though the upper arm wound was still pretty bad. I noticed that roots had been tied around his left wrist and left ankle and I asked in Pidgin why they were there. I was told that the roots came from a plant that in the "old days" was used to heal and stop bleeding. The plant is tied near the wound and its magical powers will help cure the infection. Boas had tied the roots around his son's limbs a few days after the accident. I hoped that between the roots and the sulfur powder the boy would live a full life.

I spent the rest of the day with Sege walking the beach and exploring a mangrove swamp (Fig. 3.1).

Fig. 3.1. A mangrove swamp

I was looking for mudskippers (Fig. 3.2). Mudskippers are bony fish (Osteichthyes) belonging to the genus *Periophthalmus*, and are found in swamps and estuaries of the Old World tropics in the Indo-Pacific, from Africa to Australia. Their genus name refers to their large, movable, close-set and protuberant eyes. Unlike other teleosts, they have strong, muscular pectoral fins by means of which they are able to leave the water and crawl over the mudflats and climb into the lower branches of mangrove trees. When out of water, they breathe oxygen from the water trapped in their gill chambers. We saw several of these interesting creatures in a stream about 20 yards inland from the sea.

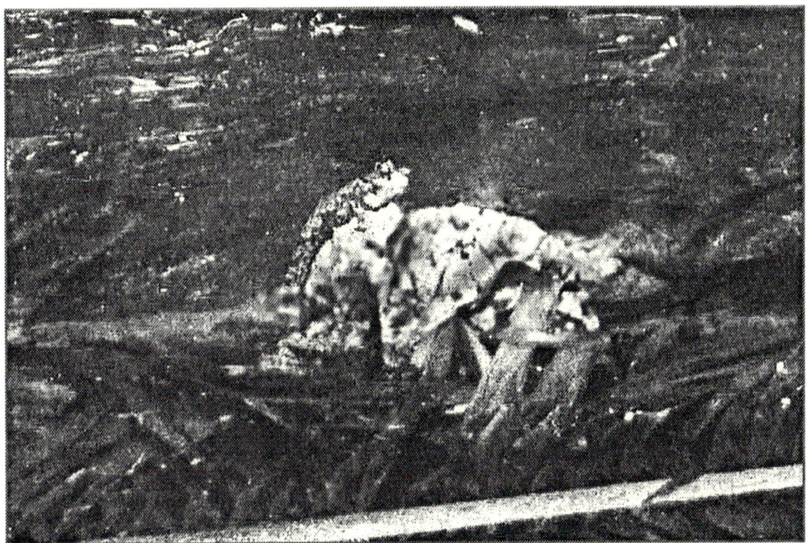

Fig. 3.2. A mudskipper resting on a stone

When walking, the mudskipper uses its ventral fins to elevate its body and move forward using its pectoral fins as legs. Its progress across the beach is sinuous and rather slow. When in the water, it uses its tail to rise up and skip across the surface of the water. The local men told me that many mudskippers live in the mangrove swamps and some are quite large and actually climb into the trees.

The next Sunday I went to the swamp and spent a couple of hours observing the mudskippers. I watched them crawl up onto the low limb of a mangrove tree and then wriggle up the branch for some distance. There they would lie motionless for awhile. I timed several of them to see how long they stayed out in the air. The average was nearly 10 minutes, after which they simply fell back into the water. It was pleasant watching them and I enjoyed the quiet of the mangrove swamp that was only disturbed by the occasional plopping of a mudskipper into the water. Indeed, Professor Loren Eiseley, Chairman of the Department of Anthropology at the University of Pennsylvania had asked me to bring one back to him if it were possible. The bottle of formaldehyde I obtained from the clinic in Talasea was for this purpose, and I hoped to get a specimen for each of us before I left the island.

April 5

My sample reached 55 by mid-morning. My work has been going along nicely and local cooperation has been excellent. I decided it was time I tried making dental impressions. It so happened that Sege's brother, Galia, was watching me measure this morning, so when I finished measuring I asked him if I could make a dental impression of his teeth. He wasn't too sure what it was I wanted to do, so I spent some time showing him the dental trays, the powder that I put in the tray, and how I would put the tray in his mouth for a few minutes and take it out again. I demonstrated the procedure on myself, hoping I wouldn't gag. I didn't and by the time I'd finished the demonstration, word had spread through the village that I was going to do something new and many people gathered to watch (Fig. 3.3).

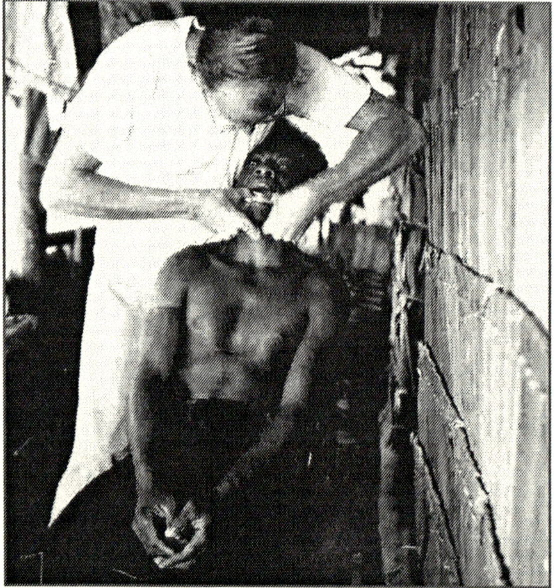

Fig. 3.3. Making a dental impression of the lower teeth

Sege's brother proved an excellent subject, remaining very still if not rigid while I placed the lower tray in his mouth (the lower tray is always put in first since there is less chance of gagging.)

People came closer and when I removed the tray, the powder had become thicker and somewhat gummy as it is supposed to. It came out without a problem and everyone smiled, several letting out a little, "Oh, oh!" Galia smiled and indicated he was ready for the upper tray. I took the upper impression without any problem, without his gagging. I secretly thanked the tooth gods, and everyone laughed as Galia stood up and smiled. After that, I immediately mixed the Albastone and poured it into the two impression trays. I shook the trays to get the air bubbles out of the Albastone as it began to harden. The procedure went well and everyone watched intently as the Albastone started to firm up and harden.

After awhile the dental casts were quite hard and I removed them from the trays. Everyone crowded around to see Galia's dental casts. He asked me if he could hold them and I said yes, but asked him to be careful not to drop them. He seemed proud and carried them around Galilo showing them to everyone he saw, telling them that they were his teeth and that I had made them. This was very positive publicity and it proved valuable. I explained to the group as best I could that I would take the dental casts back with me to America and study them. I made several other impressions during the rest of the morning, giving the rubbery, used impression material to the children who had a lot of fun playing with it.

After lunch, Sege and I cycled to Makasili and spoke to Gar, the *luluai* of the village. He told us he would have several men for me to measure in the morning and that everyone in his village knew what I did and that it was okay with them. He said that he would catch some fish tonight and that Sege and I would have lunch with him tomorrow.

When we returned I made impressions of Sege's and Uoso's teeth. Later they looked at them and said they were "No. 1," indicating high approval. I put their dental casts in a box in my room and told them where they were so they could show them to their friends. For the next several days I overheard them talking with various people telling them what I had done and showing them "their teeth."

That evening Boas came by and talked with us about the "old days" and his first memories as a young boy of the first missionaries to the Lakalai country. Ward and Ann took copious notes and were very pleased with this information. He left about 11 p.m. and we discussed his conversation before writing up our notes and going to bed. It has been a long day.

April 6

We were up early and had just finished breakfast when a man came to our house and wanted me to measure him. I did, and then left for Makasili with Sege. When we arrived Gar was waiting for us and said we could use his house for measuring the men in the village. I measured six men during the morning after which we ate lunch. The fish were cooked in coconut milk and were very good, except that their skin and scales were not removed. I removed the skin as I went along and had a fine meal.

Afterward several men showed up to be measured and, before I had finished with them, others were waiting. The talk had gotten around and it seemed they all wanted to be measured. I must say, it is a rather odd feeling to be surrounded by a large group of men in the village watching everything I did. After I finished measuring someone the "volunteer" generally would remain to explain to the next man how to stand and not to move while being measured. Sometimes it was hard not to laugh, but they were so serious in helping me that I told each one how much I appreciated his help. I asked a couple of the younger men if they were married, and several of the older men who overheard me started to rib the younger men, who were unmarried, and make jokes at their expense. Soon everyone was laughing and the younger men were shouting back insults at the older ones. It was obvious they all were enjoying themselves.

It was a nice afternoon, except for one thing: Gar's house had a tin roof and it grew extremely hot as the day wore on. Many of the houses in the villages had tin roofs, which lasted longer than the roofs made of thatch, but the houses with thatch roofs were always cooler and more comfortable. Finally I suggested we might continue in the yard and everyone agreed. I didn't want to hurt Gar's feelings, since I knew that a house with a tin roof indicated a man of wealth, but luckily he agreed with me and we continued measuring in the yard beneath several palm trees.

I left late in the afternoon and returned to Galilo where I had a swim with Ward and Ann. After dinner a man came by and showed us a polished basalt axe that had been used before the steel axes introduced by Europeans. This was the first one we had seen. After he left we went over our notes for the day and retired about 11 pm.

My spider is still there, protecting me from the mosquitoes.

April 7

After breakfast I left for Makasili to continue my measurements of the men I didn't get to yesterday. Today I decided to start from the outset working outside in the middle of the village under the trees, rather than in Gar's house with the tin roof. I measured ten men by noon and had lunch with the villagers of taro cooked with carrots in coconut milk. Taro is not my favorite food, but I managed to eat most of it before I told Gar I was full. He offered to show me his village so we took a tour around Makasili.

There are some sixty people living in Makasili. The village is similar to other Lakalai villages, with all the houses built on poles so the floor is about three or four feet above the ground. There is a men's house at one end of the village as well as several cookhouses, and the ground is kept cleared of grass and refuse by the women who sweep daily. About 4:30 pm I walked back to Galilo where I met Ann and Ward who had been with various people that day collecting kinship information (Fig. 3.4).

Fig. 3.4. Ann collecting kinship data.

April 8

Ward was not feeling well today so he went back to bed after breakfast. Ann cycled to Sharp's to get a compass to use in mapping Galilo, and I took dental impressions of two more men, bringing that sample to five.

During the afternoon, while Ann and I were away, Ward had reached for something between his cot and the wall and was struck by a scorpion. By the time we returned his hand and wrist were very swollen and he was in a lot of pain and had a slight fever. We made cold packs using the ice from our kerosene refrigerator and put them around his wrist and forearm. This seemed to relieve the pain somewhat, but he remained in bed the rest of the day and night. Scorpions are members of the Arachnid (spider) family. Their venom glands and stingers are located in their tales and although their sting can be very painful, it is not usually fatal to humans.

Later in the afternoon Sege and I walked to Vavua to see the *luluai* about measuring the men in his village. On our way back to Galilo we stopped at a group of six houses that Sege said had originally been part of Vavua, but had split from the village a few years ago because of problems among several of the men. One of the men came up to us and offered me a basalt axe. I accepted the axe and gave him several sticks of tobacco. This man has two wives that live in separate houses and he brought them to meet me. They offered me kulau to drink and we sat and talked. When he decided several years ago to move from Vavua, other families went with him and established this hamlet. This group is still friends with most of the people of Vavua but prefers to remain separate. I don't know if the trouble is related to his having two wives. There is obvious pressure from the Catholic Mission to have only one wife, but some of the older men continue to have more than one.

April 9

The rainy season is over and I'm looking forward to weather that will be warmer and more pleasant. After breakfast I measured two men from Galilo. Later Sege and I walked to Vavua and spoke to the *luluai* who told me there were about 30 men in his village I could measure tomorrow. We returned to Galilo for lunch. Ward was up and we had lunch together. The temperature was starting to climb, it was getting hot, and I didn't feel very well so I spent the rest of the day relaxing and reading.

Many times when we felt ill we had no idea what was causing it.

We had *kindam* (lobster) for dinner. These large crustaceans live in the coral reefs and we were told that they are difficult to catch. That night there was a full moon and Ward was feeling steadily better. The three of us walked down to the beach and sat in complete darkness looking at the moon. It was so restful. Ward saw a light on top of a mountain and thought it might be a signal from Father Wagner; we shortly realized, however, that it was the planet Mars rising slowly above the mountain top! We all had a good laugh at ourselves and called it a day.

April 10

I had breakfast, then cycled to Vavua to do more measuring. Vavua is perched on a knoll overlooking the beach and a small bay. It is a beautiful location and the time passed quickly. I had lunch with the villagers—fresh pig cooked over an open fire and sweet potatoes and taro mashed together in a broth of coconut milk. After lunch I worked all afternoon measuring, bringing my sample to ninety-six, then cycled back to Galilo where I met Ward and Ann for a swim before dinner. Dinner this evening consisted of lobster, fish, yams, and a salad of local greens. I thought I might lose some weight while on this expedition, but instead I believe I have gained a few pounds.

After dinner a group of people came by and invited us to come to a *singsing* in the village. We asked if we could record the music and singing, and they agreed.

In the center of Galilo a number of people, mostly young girls and boys, were already singing and laughing. They welcomed us and brought us stools to sit on while recording the festivities. One man played a *kundu* (drum), an hourglass-shaped drum with a taut skin, usually iguana, on the end that's beaten with the hand. The villagers sang and the dancers kept time with the drum by shuffling to-and-fro, stamping their feet while their arms dangled freely at their sides. It was a beautiful night, the sky was clear, and the moon was a little more than half-full. Occasionally, we would notice a young couple quietly move away from the other dancers and disappear into the shadows and into the night. As I sat watching and listening to the people dancing and singing and having such a great time, I thought of how all of this was another example of why I wanted to be an anthropologist. I mentioned earlier that Ann and I agreed that travel was a motivating factor, but there were deeper and more important reasons residing in both of us for studying such an esoteric discipline as anthropology. I had, since my junior year in High School, been interested in the evolution and the present diversity of humankind. This expedition offers me the opportunity to begin these studies. I look forward to what I hoped will be a long career.

We enjoyed the *singsing* until about 11 pm, then said our goodnights and walked back to our house. The *singsing* had been so energetic it is difficult to relax. I am the last one to retire tonight, and as I turn out the Coleman lamp I can see my spider slowly crawling across the top of the mosquito netting. I know I am safe for the night.

April 11

I was invited to go fishing today after church. After the service I cycled to Koimuimu village, a few miles north east of Galilo. It is a medium sized village located on the beach and is the last village in the southeast part of the Lakalai territory (Map 3). There I met several men and boys on the beach holding fishing spears (Fig. 3.5).

Fig. 3.5. Men and boys preparing to go fishing

I parked my bike against a coconut tree and got into the canoe, sitting in the middle with several others. Two of the men, one in the bow and one in the stern, would do the paddling. We were on our way to the mouth of the Kapeuru River, one of the larger rivers along the Hoskins Peninsula running into the Pacific from the interior of the island. The ocean was a beautiful bluish-green color this morning and low waves were helping the paddlers move the canoe along at a fairly good speed. Before long we saw another out-rigger some distance ahead of us with several men, one holding a fishing spear (Fig. 3.6).

Fig. 3.6. Spear fishermen on an outrigger

As we approached them, the water became clearer and I noticed we were passing over coral reefs. The reefs shimmered in the water and reflected a rainbow of colors. The water teemed with a myriad of sea life from a variety of shellfish to a multitude of different shapes, sizes, and colors of fish. I thought this would be where we would fish, but as I started to stand up, one of the men said we were going to the mouth of the river to spear fish, so I sat down again. Both boats continued on toward the mouth of the river, still some distance away. Before long, one of the men in the lead boat yelled that the mouth of the river was directly ahead of us and that the water would be a little rougher (Fig. 3.7).

Fig. 3.7. The mouth of the Kapeuru River

The mouth of the Kapeuru River is about 50 to 60 feet wide, and as we rowed into it, I saw what we were after: large fish swimming around the mouth of the river. They were not very deep. We spent about three hours spear fishing and keeping an eye out for crocodiles, not uncommon in the rivers along the Hoskins Peninsula.

We floated and drifted around several small mangrove islands scattered about the mouth of the river looking for fish. Whenever a fish was spotted, one person would stand up and a throw a spear. I watched carefully as the fisherman took their turns and eventually they told me to try. I stood up and, when I saw a fish near the boat, I threw the spear right at it. I missed; in fact I never did hit a fish. There was always good-natured laughing coming from the men in the boats, and I comforted myself by observing that not one of them hit a fish every time they threw.

After a while we went ashore, built a wood fire, skinned a couple of fish and roasted them intact over the fire. After we ate we went back to Koimumu where I thanked the men for taking me along and for giving me some of the fish to take back with me. We drank *kulau* and one of the men presented me with a lobster. I thanked him and slowly cycled back to Galilo with the lobster and a string of fish. When I arrived a group of kids met me and wanted to go swimming, so I gave the lobster and fish to Sege and went swimming with them. We staged a water battle which I lost.

This evening we had lobster, fish and a leafy salad for dinner. Ann and Ward thanked me for being such a great fisherman and I accepted without admitting I never speared a fish; but I believe they knew that!

It has been a beautiful, interesting day. I'm going to bed early tonight. If I don't get to sleep right away I can always watch my spider stalk mosquitoes.

April 12

I spent the afternoon measuring and, with the exception of two, completed the anthropometric survey of Vavua. I called it a day, deciding to finish it tomorrow, and had an afternoon swim while waiting for Ann and Ward to return for dinner.

We were up at 6:30, and I read for a while before returning to Vavua to measure the last two men in the village. Nothing was happening back in Galilo, so I walked to Rapuri to see Lima who, I was told, was in his garden and would not be back until late evening. When I got back to Galilo a Mr. Brown was sitting on our porch, pausing on his way to see Father Berger.

Shortly, Ann and Ward returned and we all had lunch, after which Mr. Brown went on to the Mission. Ann, Ward, and I started to make a detailed map of the village, but it was very hot so we didn't work long. After dinner, Boas came by to talk with Ann and Ward about marriage customs and bride prices.

Sege appeared a little later and announced that we were invited to a *singsing* in the village in honor of a new canoe that had just been built. We picked up cameras and recording gear and accompanied him to the *singsing* where we spent the next few hours recording songs about canoe building and the good fortune that they hoped would follow the canoe in its voyages. There were drums played by men and short hollow logs that both men and women beat with their hands or long poles. Each song referred to the canoe and was sung by both sexes. After this, the women started singing and dancing. One woman seemed to be the leader—whatever she did, the other women followed. They danced around in circles and then danced toward the on-lookers, dancing backwards again just before making contact then dancing in circles again. Once or twice they danced up to us, sang a short song then retreated. The *singsing* lasted for over an hour. As we walked back we could hear many people still singing their way to their homes.

After breakfast I left for Rapuri with my new assistant, Kalua (Fig. 3.8), who took over for Sege whose primary duties are as Uoso's second cook.

Fig. 3.8. Kalua

I have known Kalua for some time and, when I realized that Sege was doing double duty, I asked Kalua if he would like to help me and record the measurements as I took them. He said he would, that he understood my Pidgin pretty well and understood the numbers in English. As it turned out, he was an excellent assistant.

I measured six men during the morning and had lunch of tea and hard-boiled eggs with Lima who asked me about America. When I told him something of the country, the size of it, how long it would take to get there, he replied only that it was "too big and too far away." As I come to know him better I begin to understand his ambitions and those of Boas as well. I could not help but think of American politics and politicians and how the political impulse rises up in certain personalities everywhere.

On our way back to Galilo, Kalua and I talked about my work and I explained that I intended to write a book about the Lakalai. He was pleased and said he would like to see the book some day.

We received mail today, which made us happy. I had letters from my parents as well as from my wife. We will go to bed tonight thinking of friends and family at home.

April 15

After measuring all day in Rapuri my sample has reached 130 and I am feeling good about how my work has progressed.

The three of us talked after dinner and compared notes, as usual, then retired rather early.

April 16

Ann awoke with a skin irritation that made her feel unwell and kept her in bed for the day. After breakfast, Ward and I cycled the five miles to Makasili to attend the Sunday service at the Methodist Church. Mr. Brown conducted the service and afterward, we had lunch with him and some of the local people.

He told us that he first came to New Britain in 1902, but did not start a Mission until 1918. The first services were attended by Samoans and Tolai people from Rabaul. The main purpose for the Mission, he said, came from an idea held by people in Rabaul who hoped that if the Lakalai were converted to Christianity they would become less war-like. As it was, when people from Rabaul came to the Lakalai region to obtain *girigiri* (cowry shells used as money), there were many unfortunate incidents with locals along the way. Another sign of influence from Rabaul is the house style. Before that Lakalai houses were built on the ground. The present raised house type is the same as in the Rabaul region. He also talked about cannibalism in this area and which of the various islands accepted it. It was not uncommon in this region and some incidents have occurred even within the last decade. He is obviously very knowledgeable about these people and it is unfortunate that, when he dies, so will most of his knowledge of this area since he has never recorded this kind of information.

After this most enlightening discussion we returned to Galilo for our daily swim. Later, Sege offered to cut my hair and I said yes. Actually, it turned out pretty well and I thanked him. He seemed happy with the results, he said, since my hair was getting below my ears and Lakalai men wear their hair quite short around their ears.

Ward, Ann and I had dinner and retired early. I will go to sleep looking up at a full moon through the window.

April 17

Kalua came by our house after breakfast and we walked into Galilo where I took a picture of Sege sitting under his house cutting the edible, white meat from coconuts (Fig 3.9).

Fig. 3.9. Sege cutting copra

When dried, the meat is called copra and is valued for the coconut oil extracted from it. I started to take another picture of
Kalua, his wife, and their first child, a boy, and his sister standing in front of Kalua's house. His wife, however, would not let
me take the picture until after she went into their house and put
on a clean blouse and *laplap*. Vanity, I believe, is universal (Fig. 3.10).

Fig. 3.10. Kalua and family

As I walked through the village this morning, I noticed an ever-
increasing friendliness in the people. Whatever house I pass,
someone is there to smile and say hello. And, of course, there
are always children trailing behind me. Ann took a picture of
me with one of the children. The boy was not too certain about
having his picture taken (Fig. 3.11).

I returned to our house to find Ward seriously ill with an attack of malaria. I went to get Loa, the medical assistant, and when we returned Ward had a temperature of 104.8 degrees. That is getting pretty hot! We put thin ice packs from our Kerosene refrigerator on his head and after about an hour, his temperature dropped to 102.3 degrees and remained there for the rest of the day. Ward ate a little and then went back to bed.

A friend, Babo, came by not long after dinner and wanted me to go fishing with him, which I did since Ann was there to look after Ward. We were gone about two hours and, as usual, I didn't catch a fish but my friend did. We returned to the house when it started to rain. Ward was asleep and Ann was collecting vocabulary from an informant. She said Ward had not moved since going back to bed. Ann retired and I sat up talking with Babo until midnight. When he left I checked on Ward who was fast asleep. An illness like this, even when we know what it is, always makes us a little nervous, but I can feel myself falling asleep too, listening to the rain.

April 18

Ward slept through breakfast today, Easter Sunday, and seems a little better. He slept most of the day but did get up to have a little to eat at dinner, after which he went back to bed. The weather is miserable and overcast, and it has been raining steadily since early afternoon. We had thought the rainy season was over! While Ann and I were talking about the progress we were making collecting data, she casually mentioned that today was her birthday. I told her I would have baked her a cake had I known. She laughed and said that was why she had not told me. Around ten o'clock we checked on Ward who was sleeping soundly. We were tired and certainly, it had not been a *gutpela dei* (good day), Ann's birthday notwithstanding.

April 19

Ward stayed in bed the next morning though he said he felt a little better. When Kalua came by he and I gathered up my measuring instruments and started the walk along the beach toward Koimumu. This was my first day to work in this village and it always takes a little while, talking with the men, showing them the instruments, and often, having a smoke with them. During this time crowds of children would gather around me talking and laughing. Usually, as was the case today, one or two of the boys will touch my arm or my shirt (it would be extremely rare for a girl to get that close to me.) I indicated to the children that I didn't mind and continued to make my way to the house of the headman of the village. Suddenly the *luluai* appeared and began to shout at the children "*kilia, kilia*" which means "Stand back! Make way! Clear out! "

Such scenes usually occur only on the first day. The novelty wears off, the throngs thin out and, after the second day only a few people come to watch me work. Of course, by now, everyone in Lakalai knows that three Americans are living in Galilo; one of them, they have heard, is measuring people's bodies with long and short sticks.

I measured five men in the morning and then went to the *luluai's* house for lunch. We had *kulua* to drink as well as a salad of green leaves and large *kropis* (crayfish). I told the *luluai's* wife how much I enjoyed the lunch and thanked her. She smiled but did not say anything to me.

During the afternoon I measured five more men but about 3 o'clock it started to rain and I had to stop working. Within a few minutes it was a downpour and one of the men got me a hat—I believe it was an old Japanese Army hat—and another man brought me a poncho.

On the way back to Galilo Kalua cut a large banana leaf to use as an umbrella. As we walked along the beach we saw many men squatting in the bushes with banana leaves covering them while they waited out the rain.

When we got to the house Ward was sitting in bed with a hot water bottle on his right ankle. He said an infection had appeared while I was gone, and Sege and Eoso had fixed the hot water bottle for him. We kept hot plasters on the infection the rest of the evening until the swelling and redness disappeared.

While Ward was resting I had a long talk with Kalua and Sege who wanted to know how long we were staying. Additionally, they always seem interested to know if more Americans are coming. I explained to them that we would be here for several more months and that in a short while two more people, the Valentines, were arriving. They planned to stay about a year. They were pleased to know about the Valentines, as I believe they enjoy having new people in their midst.

Ann came back from working with the women in the garden and said, "Lets go swimming," so she and I swam for about an hour, and when we returned Ward was sitting up in bed and looking pretty good. We had dinner together, and for the first time in four days Ward sat with us on the porch and talked till bedtime. We are relieved that he is feeling so much better.

During our talk I told them not to be alarmed if they saw the men looking at and feeling their and other mens' wrists with their fingers. I had been examining that part of their lower arms, feeling for a muscle that lies in the middle of the wrist. The muscle is the palmaris longus and is variable in all human populations; that is, sometimes it is there, fully developed, and sometimes it isn't there at all. We know something of its mode of inheritance, and I was interested in its frequency in the Lakalai.

I told the men that some of them had this muscle and probably some did not. They became interested and I showed them how and where to feel for it in their wrists. Later, I noticed men going around looking for it in their friends' wrists. They were having a good time, particularly when they did not find the muscle. Then the examiner would laugh and exclaim, "*tispella man I no-gat mit.*" Which roughly translated is, "This man has not got it." Others would feel his wrist and they all would laugh, including the examinee.

April 20

The afternoon was quite hot and by four o'clock I had measured 14 men. We had our usual swim before dinner after which we discussed our work that day. I always enjoy these discussions since Ward and Ann keep me apprised of the local happenings and the current Galilo gossip.

After breakfast Kalua and I cycled to Koimumu for our last day's work in this village. After lunch, as we walked around the village taking pictures, I noticed a man with elephantiasis, a chronic disease characterized by the enlargement of parts of the body, especially the legs or genitals. It is caused by obstruction of the lymphatic vessels due to infestation by filarial worms. Both of his legs were extremely large below the knees and the ankles and feet were quite large and the right foot had several separate, warty growths. This was the only time I saw elephantiasis among the Lakalai. Unfortunately, I was unable to obtain any information about this person, except that he had always lived in the village (Fig. 3.12).

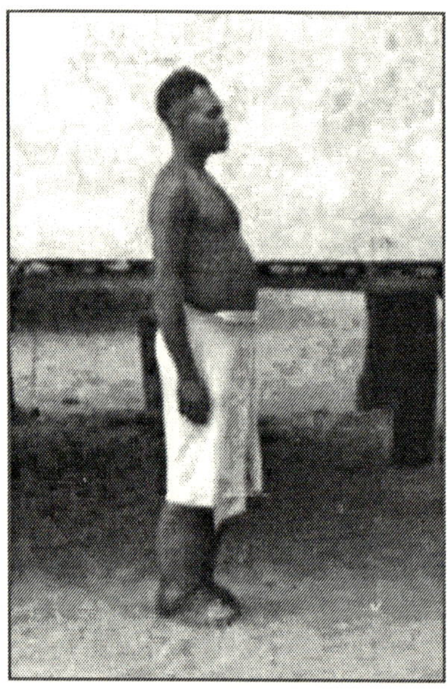

Fig. 3.12. Lakalai man with elephantiasis

As I was getting ready to start work again, another man came up to me and presented me with a small iguana lizard. The lizard is a food item here and the skin is used to cover the head of their drums. I thanked him and said I would take it back and give it to our cook. The last person I measured that afternoon was a *lapun* (older person). When he stood on the low table I used to create a level surface for measuring stature, a leg of the table broke and he fell to the ground! As I helped him up, the people watching me work started to laugh and so, after a moment, did the *lapun*; indeed, he seemed to enjoy it as much as the bystanders. He was not angry with me or the table, or indignant with his fellow villagers; it was just an accident, a joke on him. These people seem to me to be very good-natured.

April 22

Today the Valentines were to arrive from Talasea and I was to meet them at Sharp's house since neither Ann nor Ward felt very well. Ann had gone into the village this morning but returned before lunch and did not look as if she felt good. She had a fever of 102 degrees, so she went to bed. Ward also stayed home and, after lunch, I cycled to Sharps to meet and greet the Valentines.

The boat arrived from Talasea, but the Valentines were not on it! We asked the crew if they knew anything about what had happened to their passengers and we were told they would be on another boat next week. I remained at the Sharp's for a couple of hours, had tea and cookies, and discussed our forthcoming trip to visit the mountain people in the interior of the island. Mavis invited me to dinner on Sunday and I accepted. When I returned about six o'clock Ann was feeling well enough for dinner but went to bed early. I stayed up reading and swatting bugs till 11 o'clock.

April 23

I was up early this morning, had a quick breakfast, then Kalua and I walked to Karapi where we were met by the *luluai* of the village. Karapi is the second largest village in Lakalai, and the *luluai* informed me that there were about 80 men that I could measure. I thanked him and said I would measure between 20 and 25 men in Karapi tomorrow, then I would go on to measure men from all of the Lakalai villages. We walked around the village for a while and I took some pictures of children, animals and village life (Fig. 3.13).

Fig. 3.13. Domestic pigs are present in all villages

Pigs have free run in all villages and you often see people with scars on the backs of their legs where they have been nipped by them. As we left the village a group of men came up to me with a gift of a dozen or so oranges.

About mid-morning the temperature was beginning to warm up a bit so we walked more slowly and Kalua began talking about his life. He was born several years before WWII and was, there-fore, a young man during the Japanese occupation. The island was occupied about two years until the American solders came and pushed the Japanese back toward Rabaul. Later, he said, af-ter the Americans left, he was married. Unfortunately, he said, there were problems with the marriage and he left his wife. Later, he married his present wife and they now have one young boy. He said he was thinking about another wife but would wait for now. The Lakalai were originally polygamist but with the com-ing of the missionaries, about the time of World War I, polygamy had generally given way to single-wife marriages, though such multiple unions still occur.

After lunch, I thought I would make some dental impressions, but most of the men were working in the gardens so I only obtained two. At this point I have a total of seven dental casts.

April 24

Kalua and I walked to Karapi again where I measured six men. Most of the other men, I was told, were in Rapuri helping to build a new house. After lunch in Galilo, Kalua and I worked on Ann's bike trying to get it running again after an accident she'd had with a coconut tree. The rain continued to increase and an east wind began to blow quite hard. All in all, it was a miserable afternoon.

April 25

We woke up rather late since it is Sunday and we did not plan to work. About midday a group of young women and men came by and asked me if I would like to join them in a walk along the beach. I joined them and before long they left the beach and took a trail into the bush. I followed along and shortly we reached a small clearing and stopped. One of the men told me they were going to add ornamental scarification to the back of one of the girls (Fig. 3.14). A knife is used to incise the skin into which some ash is rubbed into the opening to form a keloid, an excessive growth of scar tissue on the skin. It is an old custom and today is performed in an isolated spot because the missionaries do not condone the ritual. I am pleased to have been invited to see this ritual.

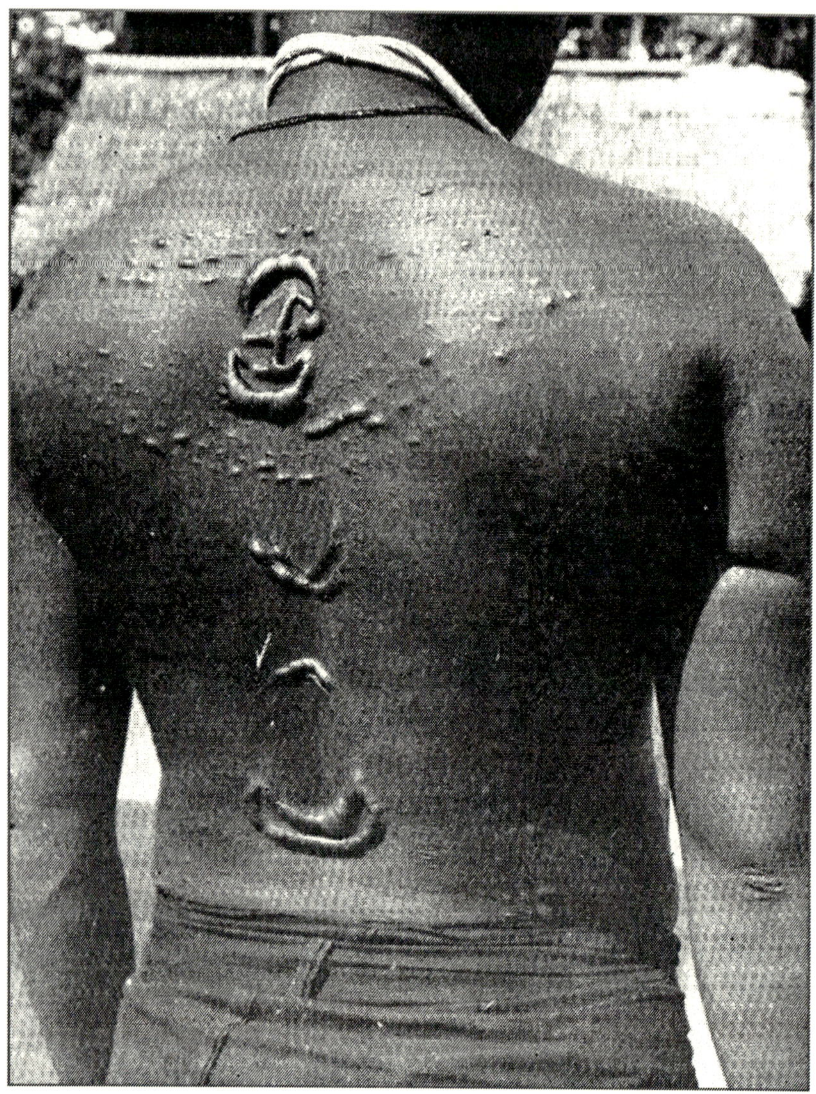

Fig. 3.14. Lakalai woman with ornamental scarification

It rained hard again this afternoon. These rains late in the day are typical for April and have been the pattern for the last several weeks.

As usual, after dinner, we talked, told jokes, when we could remember them, and retired early. My spider friend is still with me. I decided it must be getting enough to eat.

April 26

I woke up this morning with horrible stomach pains. I did not have breakfast, and when Kalua came by I asked him to go on to Rapuri and tell them that I was sick and would come there tomorrow.

By 10:30 I was feeling better, though I thought I would stay in Galilo and take some dental impressions.

My sample today increased to seventeen. I had started with impressions of young men, between the late teens and early twenties, a good age since they were not old enough to have worn down the surface of their teeth which meant I could better study the structure of the teeth. They were also better subjects than the older men, since they sat still while I was taking the impression!

April 27

Shortly after I returned from Karapi this afternoon, a man came by and told us a boat had arrived at Makasili with the Valentines. Ward said he would go and meet them though it was raining hard again and cycling would be difficult.

About 11 pm, another man came by to relay the message from Ward that everything was OK and he would be staying aboard the boat that night. He and the Valentines will be here tomorrow.

April 28

Kalua and I cycled to Gavuvu where I measured all morning. At about 10:30 I looked up to see Ward and the Valentines going by, so I called out to them and we talked awhile.

Later, when I arrived back home, Sege told me that Ward, Ann and the Valentines had gone to Rapuri. I cycled over to find the

Valentine's busily trying to get settled. They returned with us to Galilo where we all had a refreshing swim and dinner and talked until almost midnight when they left for home. It is nice to see and talk with people from the states (Fig. 3.15).

Fig. 3.15. The Valentines at home in Rapuri

April 29

I was up early today since I was going to Gule, a village about five miles from Galilo and inland from the beach at the southern base of Mt. Lolo. Kalua and I walked to the village that was mostly uphill from Galilo and built on a small knoll. The houses resemble those along the coast in that they are built on wooden poles two or three feet above the ground. I measured eight men in the morning under a large tree with spreading branches. There was a pleasant breeze blowing; otherwise, it would have been stifling.

I noticed that the people in this village were a little more apprehensive, about me and what I was doing, than were people along the coast. There were no problems, however, and by noon I had finished. The *luluai* provided me a nice lunch of taro, sweet potatoes and coconut milk. After eating, Kalua and I started back to Galilo but decided to go by Sharp's house at the Government Station. We would be leaving for the interior on the tenth of May and would be gone for about three weeks. I was invited for dinner the next Sunday and, as Mavis said she would bake a pie, I accepted without hesitation.

April 30

After breakfast Kalua and I decided to walk to the next village on my list, to Tagaragara, since I have been having hemorrhoid problems and did not want to cycle. We did not get very far until I had to return to Galilo. Sege went to the Valentines to see if they had something for pain.

I had lunch with Ward before he left for another village to collect vocabulary. Ann returned about 2 o'clock from the gardens with several stocks of sugar cane. I had never eaten sugar cane, but the children of the villages seem to enjoy it. They suck on it much as children in the States do a lollipop. I had several pieces and it was very good. I have always had a "sweet tooth," and now I had found a substitute for candy.

Ward came back about 6 o'clock with mail—the first we've received this month. He had a slight fever but said he would be ok.

There have been *singsings* each night for the last week in Galilo and Rapuri, and on several nights we heard the drums until well after midnight. I had thought about joining the party on several occasions, but my hemorrhoids have kept me close to the house. At the same time, Ann has not been feeling well and Ward has had a slight fever. Consequently, we have been party poopers.

Chapter 4

May 1

We were up early since we were to go to Rapuri for a *singsing* that would start in the morning and continue for the rest of the day. By the time we arrived in Rapuri the dinner was over, but Lima was waiting for us and said we should eat since there was so much food.

There were seven long bamboo tables, about 30 feet long, each representing one of the seven villages taking part in the *singsing*. Several pigs had been roasted, so there was a lot of meat left as well as sweet potatoes, taro, bananas, several types of fish, and green-leaf salads. After dinner the young boys sang church songs and put on a gymnastic performance followed by a series of marching formations. Lima gave a speech, and by the time we had finished eating the dancing and singing had started again and would probably last all night. We thanked Lima for inviting us to the dinner and said our good-bys to the people and walked back to Galilo.

I have noticed the past few nights that my spider has not been on the mosquito netting. I fear it is gone, perhaps to find better hunting somewhere else.

May 2

After church I cycled to Sharp's for a visit and dinner. I arrived about noon and Ernest greeted me with a glass of wine. Dinner was a meat pie, sweet potatoes, taro, and fresh green beans followed by a dessert of papaya and baked cookies with vanilla icing and raisins. I told them I would have to rest awhile before I could cycle back to Galilo!

We discussed our forthcoming trip into the interior of the island. Ernest had made similar trips before and knew what to expect regarding the jungle, swamps and high mountains we would cross. The people, he knew, were generally friendly as he had visited most of the villages before. On his last visit, however, he'd learned of some villages in a region he had never been to and he planned to include them this time. The main reason for the trip is to take a census of the people living in this part of the island. Also, he has to look into the matter of the murder he'd told me about earlier. If the murderer is found, he will have to be brought back with us and turned over to the authorities in Talasea.

Before leaving, Mavis made tea and I had a few more of those wonderful cookies. I thanked them and left in the late afternoon. Needless to say I am quite excited about the forthcoming trip. I thought about it all the way back to Galilo.

May 3

The weather is now delightful. The rain is over, I think. Temperature ranges between 75 and 85 degrees F., the nights cool down slightly and there is usually a pleasant breeze blowing in from the ocean. This is what I always thought the South Seas would be like.

After breakfast I went into the village to make dental impressions. I now have 27 casts and have used about half of my dental supplies. If I continue at this rate, I should have between 50 to 55 dental casts by the time I leave.

After dinner we had visitors from the Central Lakalai region who were on their way to visit Talasea. It was an interesting evening and they talked a great deal about "the recent war" (World War II). I saw no obvious physical differences between them and the West Lakalai. I became even more interested than ever in my trip to the interior of the island since Sharp has told me he thought there were some physical differences between the people of the interior and the coastal populations.

After they left we talked for a while and then retired.

My spider has not returned.

May 4

Kalua came by after breakfast and we cycled to Gule, one of the Lakalai villages that was not along the coast (Map 3). We were cycling along a rather narrow jungle path when all of a sudden a *muruk* (cassowary) sprang out of the bush and ran across the path in front of us! It startled us so that I nearly fell off my bike. Quickly it disappeared into the bush, and Kalua and I stood still to see if we could hear it. It was probably as frightened as we were, and we did not see it again. As we rested with a drink of water, Kalua told me how dangerous these birds are. They can kill a pig, and he said he has heard that they can kill a man with their powerful legs. They have one long claw on the rear side of each leg that they use as a dagger when fighting. In fact, the Lakalai in the past would use the claw to tip their spears. Kalua also told me the natives were more scared of the *muruk* than they were of wild pigs. This was my first encounter with these birds. Up to this point, I only had eaten their eggs.

When we arrived in Gule we were told that the men were at Ri-kau, attending a *singsing* for an important man who had died; consequently, there was no one to measure. We rested awhile, had a cooling drink of *kulau* and started back to Galilo. On the way back, Kalua told me a story about a man he knew in Ri-kua who had gone to New Ireland to live a few years ago and married a girl there. Every thing was fine until he had an af-fair with a young daughter of a *luluai* who became very angry. Somehow, it is believed, the *luluai* was able to obtain something that belonged to Kalua's friend or that was part of him, e.g., hair or fingernail clippings, and cast a spell, for within a few weeks Kalua's friend had died. The *luluai* was suspected of being a sor-cerer, able to cast such a *poisin* (an evil spell). Such stories are common among the Lakalai; in fact, the belief in the power of sorcery is common throughout Melanesia and may itself be the most important factor in the efficacy of sorcery. The murderer that Sharp will be looking for on our trip into the interior of the island is reported to be a sorcerer. The trip becomes more intriguing all the time!

It rained a little in the evening, but we hope the monsoons are over. I imagine that when I get to bed I will dream about sorcerers.

May 5

Today turned out to be an exceptional and rewarding day for me. After breakfast, I noticed several women walking toward the house. I got up to greet them and the woman in the lead exclaimed, rather nervously, that they had come to be measured. This came as a complete and gratifying surprise to me. I had never tried to persuade the women to be measured, although some of their husbands I spoke with said it was OK with them though it would be the wives' decision. I had more or less decided to forget about it. When Kalua arrived to do the recording he seemed as surprised as I was.

The first woman to step forward seemed a little nervous when I touched her arm. I asked her to stand straight and still, and demonstrated the position for her. Everything went smoothly. When I finished measuring the first woman, another one came forward. Several husbands were watching rather soberly but appeared to have no objections and they went away. The women began to talk and laugh among themselves, and all was fine after that. I measured five women during the morning and seven more in the afternoon. The women not measured said they would be back tomorrow morning. Kalua told me they were having a good time, and I could probably measure as many women as I needed for the study.

May 6

We were up early this morning to go wild pig hunting with some of the men of the village. They hunt wild pigs with large nets made from the inner bark of a large tree. A net is about five feet wide and may be as long as sixty feet. The nets are kept in the smoke house, so bugs and mold cannot destroy them. They belong to a particular sib or descent group, though most of them seem to be jointly owned and cared for since men from different sibs make different sections of them. The ownership is quite complicated.

In hunting, the largest net is carried into place in the middle of an arc by men standing about five feet apart along the length of the arc. Only men who have not had sexual relations for a week are allowed to carry the net. Other men stand outside the net carrying long spears with sharp iron tips that in the old days would have been tipped with the claw of a cassowary bird. These men will not use their spears, however, unless a very large pig is trapped or it becomes necessary to protect the men carrying the net.

Before the net carriers leave the village, men called "beaters" go out ahead of them into the jungle for some distance, then turn and walk slowly back toward the net bearers, making as much noise as they can. If there is a pig in the area, it will run away from the noise toward the net. If it is a large pig, it will be speared while in the net; if it is small, they will use the net to hold the pig down and smother it. Afterward it is tied to a pole for carrying it to the village. As they bring the pig into the village the men will sing a song announcing they are bringing in a pig. People come out to meet them at the edge of the village and there follows much singing and dancing. Unfortunately, we were all disappointed since we did not find a pig that day.

It rained again late in the afternoon, but Ward and I went swimming anyway. I must say, I am not too sure about the so-called rainy season and not-rainy season.

Ann is still doctoring her skin ulcers with sulfur power and applying penicillin directly on them.

After dinner Ward showed us how to make caramel candy by boiling a can of condensed milk for about two hours—it is delicious.

At 3:30 today it started to rain and continued for over two hours. I asked Kalua about all of this rain. He told me that it would rain off and on for the next couple of months and usually it was only in July and August that it did not rain. It seems that the rain lasts from December through March or April, and even into May or June. There is more rain than I expected—according to Kalua it sometimes rains every month of the year. The table below shows average rainfall over a seventeen-year period as compiled by the government Weather Station in Talasea on the Willaumez Peninsula.

Table 2. Average Rainfall in Talasea on the Willaumez Peninsula (for 17 years) and Malalia on the Hoskins Peninsula (for 3.6 years) in inches.

MONTH	TALASEA	MALALIA
January	33	23
February	27	25
March	24	30
April	20	15
May	8	8
June	5	6
July	5	5
August	5	6
September	4	5
October	7	5
November	10	13
December	20	10
TOTAL	**168**	**151**

At this time of year the rain usually comes in a torrent for about half an hour then tapers off for another hour or so and stops. As we have been witnessing this month, there is often also thunder, lightening, and high wind. It never gets boring, since the wind shifts and we are never sure of the direction it is coming from: first one, then another.

There is usually a quiet period everyday about 5 pm when the wind stops and mosquitoes and dozens of other insect species begin to swarm. This can last for two or three hours during which it is not unusual for us to take to the protection of our mosquito netting. The natives call this period the time of the Southeastern Winds. One gets use to it in time, they tell us. We learned to keep on working and not think about the bugs.

I must comment about the social custom of chewing the nut of the areca palm, *bilinat* (betel nut), common throughout this region as well as much of Southeast Asia. I chewed my first betel nut following the directions of Kalua. First, he told me, bite the outer husk off the betel nut. Inside is the nut itself (about the size of a US quarter.) Then add lime produced by burning wood and seashells together, wrapping the nut with the lime in a leaf of *Piper betle* (betel pepper plant.) Once the wrap is in your mouth, let it mix with your saliva and either chew or simply let it rest against your cheek (if you let it rest, after awhile you will notice a slight warming on your cheek). During this time, if you spit, your saliva will be colored red because of a chemical reaction between the betel and lime. The Lakalai chew betel; children, women and men, everyone chews betel. They say it makes them strong and happy. They have an expression "*bilinat bolong mi oleman, wiski belong yu oleman.*" Roughly translated, this means "betel belongs to my people, whisky belongs to your people." At times I did notice a warm spot on the inside of my cheek, but it was not alarming and I never had a feeling of being "high" chewing betel.

May 8

When Ann came to breakfast this morning we noticed that her face was so swollen that her eyes were almost closed. Her whole body, she said, was covered with a rash. Ward took her to Makasili to see the medical assistant and neither have returned yet. It is raining, thundering, and there is much lightning tonight. It is very dramatic and matches my mood as I am worried about Ann.

May 9

This morning, Sunday, the church bells were ringing so loudly they woke me up. I bolted out of bed before I was completely awake, looked around and wondered where everyone was. Then I remembered that Ann was ill and Ward had taken her to Makasili. When he returned Ward told me Ann would go to Bell's hospital in Talasea tomorrow. We put some of her things in her suitcase and cycled to Makasili to see her. In addition to the swelling, she has a cold. Her eyes are swollen almost closed, and although she is totally miserable, she said she would be OK and wished me well on my trip tomorrow to the interior.

Ward and I went back to Galilo and had dinner with the Valentines. After dinner, we had a few drinks celebrating my departure and toasting Ann's speedy recovery. I have been looking forward to the trip with Sharp for a long time and do not think I will be able to sleep with all this anticipation.

May 10

I woke early, had breakfast, and sat on the porch waiting for Sharp. At 9 o' clock he came by with eight *polis boi* (Native Policemen) from the Talasea region. On such trips, he told me, he always uses policemen from other regions rather than those who may have come from the areas they are visiting. He reminded me to keep an eye out for wild pigs and cassowaries. I had already seen how dangerous wild pigs could be. We had our tea and were off to pick up our first carriers in Koimumu.

I was surprised that many of our carriers were women. I was told privately that women were considered more reliable and trustworthy than men and better carriers. They would customarily stay with us for one or, perhaps, two villages, depending on the distances between the villages, Sharp told me. When new carriers joined us, the previous carriers returned to their own village. In general, he explained, people did not like to travel too far from their own territory so it is standard procedure to hire one series of carriers after another along the way. I could not help but be reminded of old Tarzan movies I had seen years ago and the problems encountered, no doubt exaggerated, when native carriers ventured too far into foreign lands.

We left Koimumu and headed inland along a rather rough trail thick with underbrush, vines, and exposed roots. The jungle became more dense with flowing plants and ferns as we walked along; the trees taller and more resplendent with mosses, lichens, fungi and hanging vines and flowers; the path more difficult to follow. *Kunai* (tall, tough, long-bladed grass) became more common. Sharp and I went ahead with a couple of the policemen and used our machetes to cut a wider path for the carriers. After about an hour or more of this, the path widened again and the terrain became less bushy. There were fewer trees. We had come to the "hot area," a very interesting thermal region with bubbling springs and steam vents, a miniature Yellow Stone National Park: the "egg fields" I had heard about, an open area lying to the south of the coastal Lakalai country and not too far from the volcanoes, Mt. Pago and Mt. Pagovah (see Map 3 and Fig. 4.1).

Fig. 4.1. Thermal area with hot springs and steam vents

This is where the bush fowl, megapodes, lay their eggs underground to incubate in the natural heat of the ground. The local natives go to the thermal grounds to collect bush fowl eggs, and use the hot springs for boiling the eggs and, occasionally, for roasting pigs. In times past, there have been fights among the various tribes as to the rights to this area, but most land use issues have now been solved. The region is about a quarter of a mile in circumference and requires careful walking since the megapodes have dug many holes. It is into these holes that the women have to descend in order to collect the eggs.

We arrived in Lavegi, the first non-Lakalai village I have visited. It is not unlike the Lakalai villages along the coast. The houses are built on poles, the grounds are clean, and the men's houses are not raised but built on the ground at the ends of the village. Here Sharp paid the carriers who returned to Koimumu. We would get new carriers from the village before we left tomorrow.

Later in the afternoon the men of the village put on a sham battle in our honor. Two opposing groups, painted with white paint and carrying shields and spears, approached each other from opposite ends of the village. As they advanced, they yelled and thrust their spears at each other and danced from side to side, all the time getting closer but stopping just before they reached each other. At some point one of the elders called an end to the show, and it was over. It was an interesting demonstration of how villagers fought one another years ago. One of the men agreed afterward to let me take a picture of him holding his shield and spear (Fig. 4.2).

Fig. 4.2. Man holding a spear and shield

That evening we had chicken and sweet potatoes for dinner and afterward sat with some of the men of the village, talking about our trip and chewing betel nut together. Everyone spoke Pidgin. Although I did not measure them, I was sure that they were similar in size, skin color, and hair form to the Lakalai.

After dinner, a man came by and told us he would go to the next village to let them know we were coming. Sharp told me it is a good idea to have a man go ahead to the next village so as not surprise a village with our sudden appearance. Before we went to bed, a man came by and presented me with a shield and a 12 foot spear. The shield is made of a hard wood and the spear of black palm. Both are decorated with bird feathers and cowry-shells. I thanked him very much but told him I could not accept them since it was going to be a long trip and I would not be able to carry them all that way. I hoped I had not insulted him. Sharp said it was OK since the man now understood that we were not going directly back to Lakalai.

Tonight I am physically tired, but my mind is still active with all I have seen and heard.

May 11

After breakfast, Sharp had the villagers line up for the census while I took some pictures of the village and people. Afterward, Sharp arranged for carriers to take our goods to the next village (Fig. 4.3).

Fig. 4.3. Carriers preparing our bundles of supplies for the trip

We left about 9:30 and entered the rain forest jungle where it was dark, hot, and humid. Many of the plants were taller than we were and the temperature gradually climbed to between 95° and 100° F. We walked along, the path becoming more crowded by the tall trees and dense undergrowth until we finally broke into an open area at the edge of a river. It was a narrow and rather swift river, but a nearby village provided canoes to carry us across. From there we had a steep climb up the side of a mountain to the crest from which the trail descended into a thick jungle and then back up the side of another mountain. We encountered this type of terrain for the rest of the day: up one mountain, down into a valley and up another mountain. The mountains were not very high but were so steep that often I had to "climb" by crawling up on all fours. The carriers, mostly women, went straight up, carrying their loads on their backs. They were amazing.

It was hard for me to believe, but here I was, in the middle of a jungle in New Britain, with a group of people that I was completely dependent upon for my survival. At last, I thought, I am an anthropologist.

We shoved off again, slowly descending a narrow, steep path that took us to a valley through which a narrow stream serpentined until it disappeared into the jungle. At 4 o'clock we rested for half an hour. The water was cool, clear and quite refreshing and Sharp said we it was safe to drink it since we were near the headwaters of the river. We also soaked our feet and several of the men went swimming. We relaxed, drank *kulau*, had a smoke, and prepared for the next walk. We had crossed the narrow valley and knew the climb up the next mountain would be arduous. The mountains in this region of New Britain are difficult to navigate and the trails sometimes hard to find. As you climb, if you look up, it seems as though the person ahead of you is practically standing on your shoulders. The soil is slippery, roots crisscross the ground, and all in all, it is damn hard going even for a so-called "mountaineer" from West Virginia.

By about 6:30 it started to get dark. Night approaches quickly in the interior, but Sharp told us we still had about fifteen miles to go. We rested a bit and as the moon rose, we lit lanterns and were off again.

Between the moon and the lanterns there was a little light, but the going was difficult and dangerous, especially for the carriers. We had to trudge on, though, since we had no tents and spending a night in the jungle without them would not be pleasant. We arranged the lanterns as follows: one in front, the second in the middle, the third at the end of the line. It was slow going and the climb seemed to get only steeper and darker, but eventually we made it to the crest of the mountain and could see the lights of the village. It was still a half-mile walk to the village and by the time we arrived it was 9:30. We had walked for almost twelve hours, the last four hours in the dark. Villagers came out to meet us and were very friendly. All was well. Sharp and I snuck away to the house provided for us, for quick, cold showers under a bucket and made ourselves presentable for dinner.

I can't remember when I have been more exhausted. I have blisters on both feet and virtually every muscle in my body is aching. Worse, my hemorrhoids have flared up and are beginning to hurt. I really don't know how I managed the last few miles except to remind myself that I had no choice.

I must add a word about the magnificent men and women who carried the supplies. Their strength and stamina are unbelievable. Several of the boxes weigh over 80 pounds and are carried by two carriers. Smaller boxes, about 50 pounds each, are carried by a single person. One man, I saw, carried a bag of rice weighing well over 80 pounds for almost three and a half hours before putting it down, and then only to rest it against a tree a couple of times.

After it became dark, when the going was especially difficult, the carriers would occasionally break out in song or, if they weren't singing would just shout out to one another or laugh. Their spirits were consistently good.

Today we'd climbed and descended eight mountains and crossed six streams. The terrain in this region reminded me of a magnified washboard with low but steep ridges with narrow gutters between. Usually streams meander through the valleys but sometimes we found nothing but rocky beds where the water had dried up. Sharp figured we walked between 20 and 25 miles that day, mostly in an inclined position, either going up or going down a mountain.

The village we stopped in was Umoa, located on what a geologist refers to as a spur, or hog back, extending along the crest of the mountain (Fig. 4.4).

Fig. 4.4. Umoa village stretching out along a mountain spur

Sharp told me that many of the interior villages were originally located out on these mountain spurs for protection from other villages. After climbing up here, I understood why this was so effective. When we sat down to dinner that night my legs started to quiver, shake, and roll. I was still climbing mountains. We ate boiled rice, Bully Beef, canned fruit, and several cups of cocoa. After dinner Sharp asked me how I was doing. I told him I was fine, though my experience in the Navy in World War II had prepared me to sail, not walk, to our destinations. We laughed and had a couple shots of rum and talked some more. He told me that today was the worst leg of the trip. I replied that was fine with me. The next village we were going too was Ainbul, located on the Gasmata side of the island—only about a two-hour trip.

May 12

We were up early for breakfast but decided to stay here today and rest. Umoa is approximately 1,500 ft. above sea level and has a population of about 90. Most houses are built on the ground. There is a three or four-foot high fence around the village for protection from wild pigs or surprise attacks from other villages, although the latter are rare today. The ground within the village is clean, and the houses are well constructed of thatch. There has been frequent contact with foreigners through the years. As with the Lakalai, these people are matrilineal and have gardens where most of their food is grown, though they also eat fish they catch in the rivers. Their spears and shields are highly decorated as they are among the Lakalai. The shields we saw are painted white, red and black. We were told that these items are passed through the male line from generation to generation.

We intended to spend a quiet day visiting with the people, but Sharp located the "sorcerer" we had come for, and held court. The man freely admitted to killing a man who had insulted him. He claimed to have powerful magic and was not sorry he had done it.

Sharp sentenced him to three months in jail in Talasea. Later, Sharp told me privately, the man would be put to work in Talesea and never really be confined to jail.

We obtained new carriers and left after breakfast for Ubai. The trip was again difficult because of the many mountains and valleys to traverse. About half way there, we were met by a group of people from the village of Lavegi (Fig. 4.5).

Fig. 4.5. Ernest Sharp and native police talking with people from Lavegi village

We rested awhile and the Lavegi people accompanied us to Ubai where we arrived around 4: 30 in the afternoon. My legs and feet were hurting and my blisters were no better. We had our usual Bully Beef, rice, sweet potatoes, and canned fruit for dinner. Sharp and I talked after dinner. He was worried about my feet. I said I was OK, and after a couple of drinks of rum I did feel much better.

It is much colder at night in the mountains than on the beach.

Our group sleeps under blankets close to a fire that burns through the night.

May 14

While Sharp was conducting the census I walked around observing and noted that physically (skin color, hair color and texture, and height) the people are similar to the Lakalai and to the other inland people I have seen on this trip so far.

This is a rather large village, about 208 houses. After census, we left for Kukula, a village we could reach in a three-hour walk. After climbing a few more mountains and traversing a swamp and five rivers, one of which I fell into, we ended up in Kukula about 4:30 in the afternoon. On the way we had passed through an area of the tallest and most beautiful trees I have yet seen on the island. I am sure many of them were well over a 100 feet tall.

This village is situated on a sloping mountain crest. The 53 inhabitants are Methodist. The houses are built on the ground, there is a fence around the village and, as with all the people we have visited since leaving the coast, speak the same dialect, or as they say, "people one talk."

My feet and legs are feeling much better. Shortly after dinner a beautiful half-moon appeared over the ridge of a mountain. It is a clear, cool night and we will each need a blanket against the mountain air.

May 15

Up early, we had tea and scones. By then the people of the village were beginning to line up for the census (Fig. 4.6). The *luluai* informed me that the people of these villages "mix" with or find spouses within the various villages in the region of "people one talk." This supports what the *luluai* of Lavegi told me, which is that the inland people mix very little with the coastal people.

Fig. 4.6. Sharp taking census in Kukula

We left Kukula at 9:30 and arrived at Mirapo about 12:30. The trip this time was not too rough. It took us through a small swamp, and around the base of several hills before we had to climb a larger hill. Mirapo (Fig. 4.7) is situated on a spur projecting outward from the side of the hill.

Fig. 4.7. Mirapo Village in the clear between ranges

As is the case with all the villages we have been to, there is a lot of evidence of foreign contact: axes, iron pots, cooking ware, and metalknives,leather belts, trade beads, etc., just to mention a few items.

A census was taken of the people of this village as well as those from Roko who had come down from a mountain to the south to participate. Sharp was a little surprised by this addition since he had only encountered the Roko people one other time during his several trips in this region. Again, I saw no major physical differences between the Roko people and the other mountain people I have seen so far, though I noticed some slight differences: men in the mountains do not appear to be as muscular as the men from the coast, e.g., their legs are skinnier, and their chests are not as muscular. Their faces are more "oval" and there is little facial prognathism, that is, in profile the lower part of their face does not project forward. Skin color is similar, as is hair color. The texture of the hair is the same. Male body hair and head hair patterns are similar here and along the coast; head hair is frizzly, curly, and straight. In my opinion, many hair types depend as much on cosmetics as genetics. The shape of the nose is straight, rather narrow or wide, and the nostrils flare or do not flare. These different shapes are also found among the Lakalai. Of course, since I have not quantified these observations they mean very little scientifically. There is so much to be done in this area of the Pacific before the information is lost forever. I wonder if there is time enough to do it.

We had dinner, a couple of rums, and now it is time to sleep.

May 16

The next day we were off to Kai, a trip that promised not to be too strenuous for which I am grateful as I am beginning to have early symptoms of hemorrhoids. We crossed a small swamp and forded the Luvi River where it was about 200 feet wide. The water was low and the bottom rocky, so Sharp and I took off our long pants and waded across the river in our underwear. The water was very cold but refreshing, and as we inched our way across the river the carriers all had a good laugh at our expense. We joined in the laughter.

After we crossed the river the trail was pretty level and the mountains lower and not so precipitous. We walked along a path in the valley, eventually coming to another river. This one was about 25 feet wide and crystal clear. We removed our shoes, socks, and pants again to cross. After dressing again we walked the short distance into Kai. Census took about two hours after which we left for the next village, Gaikeke, arriving at around 2:45 in the afternoon.

The houses in this village are constructed on poles as in Laka-lai, and the village interior is much like the coastal villages. The people dress the same as the coastal people, and clearly there has been a lot of contact between the two populations. Here we saw our first dry rice cultivation, that is, rice not grown in paddies. Sharp told me that these people grow it in small amounts and, as far as he knows, this is the only place that grows dry rice. The crops here are similar to those in Lakalai, though the stable crop is the sweet potato rather than taro.

In all of the villages I have visited on this trip so far, I have noticed that the people are somewhat shy and a little apprehensive of me. They appear basically friendly, however, and I have the feeling that if I could spend some time with them I could get their confidence.

Tomorrow we climb more mountains, according to Sharp. I can hardly wait, so I retire at 9: 30.

After breakfast we left with a new group of carriers for Umu. The trail cut through high grass for a while and then there it was: another mountain directly in front. We rested a bit when we reached the top and then descended into a narrow valley that separated us from a second, slightly higher mountain. Umu was on top of the second mountain.

As we climbed, Sharp told me the following story. In the 1920's the Australian government sent an expedition to disperse the Central Lakalai who were having trouble with gold miners in their territory. When the expedition arrived the native people scattered, but during the ensuing fighting four miners were killed. The *luluai* of Umu was instrumental in the killing of the four white gold miners. While trying to flee, he was shot in the shoulder by one of the native policemen, but he managed to get away. He spent the night in the jungle then made his way to Umu. None of the natives were captured and arrested, but the current *luluai* of Umu claims to be one of the natives involved in the incident. The villages have been peaceful ever since.

When we arrived in Umu, we were met by the *luluai*, an old, wrinkled man. He and Sharp talked a few minutes and then I was introduced.

While Sharp took the census of the village, the *luluai* and I sat under a tree and talked. He also told me the story about the incident with the miners. It was cool sitting under the tree, and I enjoyed listening to him recount what was probably the most important event in his life, even if he did embellish it here and there. I asked him if I could take his picture and he agreed (Fig. 4.8).

Later we said good-by to the *luluai* and left Umu about 1 o'clock. Sharp asked me if I had understood the old man, and I replied, "Some of it." Hearing the story beforehand certainly had helped me.

The trail was clear and we had an easy walk to Vasilau. Arriving about 2 o'clock in pouring rain, we immediately took shelter in one of the houses. The village is a small hamlet on the side of the mountain that looks down into a deep valley. The houses are built on the ground and rather close together. This is the second village we have seen with a stockade around it (Fig. 4.9).

Fig. 4.9. Vasilau Village with stockade

As mentioned earlier, in times past such stockades were built to keep enemies from having easy access to the village and to keep out wild pigs. Today, they are just used to keep out the wild pigs. There were two entrances with gates, much like stiles, at either end of the village. After Sharp took the census we moved on to another village, another three hours walk.

Sipa is on another, slightly lower mountain than Vasilau and the walk was not too difficult. We stayed in the house *kiap*, a house built for patrol officers, outside the village on a spur of cleared ground that overlooked the verdant jungle stretching for miles below us. We had a magnificent view of the sunset, and to add to the affect, tonight there was a full moon. After showering, Sharp went into the tent and lay down to rest. Suddenly the supporting poles of the cot broke! He hit the ground with a noticeable thud that brought the attention of the policemen who ran into the house and asked in Pidgin, "*stik im bruk?*" ("did the stick break?") When he heard those words Sharp could only laugh and say, "*nogat, mi laikim daun hia.*" ("No, I like it down here.") Everyone laughed and it became a joke whispered for days among the police guards.

May 18

We were up early, preparing our gear and hiring new carriers. Sharp told me that from here we would have about ten days of mountain climbing into the Whiteman Range which has peaks well over a mile high.

Sharp took the census during the afternoon. Before coming on this trip I heard that the people in the interior of the island were shorter (pygmies) and that the men had more hair. I do have the feeling that some of these people are slightly shorter in stature than the coastal Lakalai. Their faces are shorter, and there is somewhat more facial prognathism, but certainly they are not pygmies. Of course, this is only a visual observation. I wish I had more time to spend here, to do measurements and take dental casts, but that is impossible.

May 19

We left Sipa about 8 o'clock traveling along a steep, curving path for about a half-mile to the Evill River. From there we crossed the river and headed upward along its west side on terrain that gradually became steeper and steeper. By 10:00 o'clock we had arrived at Elobe, a village situated on a flat spur about half way up and extending north from the mountain with a beautiful view of the jungle below, Mt. Lolo to the west, and Commodore Bay to the northeast. When we entered the village they presented us with a rack of bananas and several coconuts but would not take the trade tobacco we offered in return.

The village is probably about 4, 000 feet up the side of the mountain. The houses are built on the ground, and the interior of the village is clean and rectangular with an entrance at either end. There are a few coconut and betel trees (Fig. 4.10).

Fig. 4.10. The interior of Elobe Village

Another village, Mulusi, is about a half hour's walk from Elobe, so Sharp asked that the Mulusi people come to Elobe for the census. About an hour later they arrived and Sharp took census of both villages. Afterward we went to the house provided for us, and shortly people began to come by to visit. One of the men brought a spear adorned with cowry shells and feathers which Sharp bought from him. The man told us, incidentally, that four or five spears plus a pig would be the price for a bride.

The children were, of course, curious about us but at the same time a little suspicious since many of the younger ones had never seen white people. They grouped around us, felt our clothes and touched our skin. After lunch I was walking through the village and one of the boys, perhaps six or seven, came up to me and took my hand and walked through the village with me. We talked a little in Pidgin, and he would occasionally squeeze my hand and smile at me. He stayed with me for more than an hour.

The *luluai* came by our house and told us there was a small *singsing* in our honor that night. After dinner we went to the center of the village where the people were gathering, many decorated in feathers and leaves and painted with white ashes and lime, each carrying branches from various trees. One man beat the rhythm on a drum with a stick while a woman blew a conch shell. Many of the people started to dance around the center of the village as singles, then in pairs, and finally in trios, always moving six to eight steps forward and then three backward and continuing in this pattern till the end of a song. After a couple of songs, they danced toward us and motioned for us to get up, so we got up and became part of the dance. The rhythms were similar to those of the Lakalai, and the people told us they knew some of the Lakalai songs. The *singsing* lasted for about three hours. We danced several times and the people were obviously happy that we joined in. We went back to our house and went to bed.

I am asleep almost before I finish writing this. It has been a glorious night.

May 20

We said our goodbyes and left after breakfast, descending a narrow, steep path to the Evill River again and following it a short distance before climbing our first mountain of the day. The trail was winding and steep all the way to the top. A short distance up the mountain it started to rain and the higher we climbed the harder it rained and the denser the fog became. The top of the mountain is 4,500 feet and was just in front of us when Sharp shouted "quake!" The ground was moving under my feet and I noticed the trees weaving and shaking. We picked up the pace and hurried down the other side of the mountain, not stopping until we reached our next village, Youyau, about 10:00 am.

Sharp lined the people up for census in the pouring rain. There were no complaints. I think the people were glad to get done with it, and soon we left for Lavagi, a short walk for which I was grateful.

Apparently, before World War II, there had been one large village here, but after the war friction among the people caused Lavagi to split off from the larger village and move to its present location where it now consists of ten houses all built on the ground in a circle. For the first time I saw several men with their hair over their shoulders. Additionally, several women had light brown skin.

It was raining hard so we did not stay long but went on down to a river in the valley and up a very steep mountain to the next village. The trail was steep and slippery, and several times I had to get on all fours and crawl for several feet. Again, I am amazed at the women carriers. They are bare-footed and dig their toes into the soft dirt (mud) as they climb, allowing them to stay almost upright. I thought that perhaps at the next mountain I would remove my shoes to see if that helped me, but I never did.

We arrived in Lavugi just after noon in a pouring rain. The temperature was only about 50 degrees F. After warm water showers from a bucket, we dressed and put on jackets. Lavugi is situated along a mountain spur that extends from near the summit of the mountain outward toward the west for some distance. The mountain range is more than half way across the north-south width of New Britain. From it, the rivers flow onto the south coast. The houses are built on the ground along the length of the spur. The interior is clear of shrubs and trees and kept clean by the women. The men's houses are at the ends of the village—long, rectangular buildings similar to those we had seen in other villages.

Later that afternoon, as we sat talking with some of the villagers, we felt the ground moving and shaking again. It was not long or violent, and later the natives told us that these quakes happen "every now and then." I wondered, *how often is 'now and then'?* After the quake and rain were over, Sharp called for the census, and people lined up in front of a table behind which Sharp sat with pencil and paper. I observed them as they passed by and did not notice any physical differences in these people from the other mountain villagers, or for that matter, from the coastal people. The major difference that I noticed was not physical but behavioral. Mountain people seem to be more shy, timid and nervous than the coastal Lakalai in the presence of "foreigners," perhaps due to the fact that coastal people have more frequent contact with the world beyond the island.

Sometime after Sharp had taken census the *luluai* and several elders came by to talk with him. They informed him that they had bought land from the people near the village of Wolo and wanted permission to move their village to this new land where they could grow coffee and cocoa for export. Several said that they had worked on plantations and understood how to plant and harvest these crops. Sharp said he understood and told them he would meet with them in Wolo on his return and discuss this more at that time. Later he told me that the government was not against the movement of villages, but had to exercise some control since a move might inflame old rivalries that could result in *pait* (fighting battles) among villages. Many high villages he said wanted to move to lower altitudes in order to grow a wider variety of crops and have more contact with the outside world.

As we ate dinner a fog began to settle over the mountains and it was getting much cooler. Tonight, even with two wool blankets, it is damp and chilly.

May 21

We got up at 6 am and went to sit by the fire the natives had built outside the house. We were still in our blankets, and the dense fog hung over the mountains. There was a heavy mist on the trees, bushes, and us. The fire felt good, and we stayed there until breakfast, after which we left along a trail not unlike those we have been following since entering in the mountains.

We reached the village of Kaiko about noon. For the last several days we had been traveling through a very rugged terrain. I looked back and down, at the series of sharp mountain peaks and deep narrow valleys between them (Fig. 4.11). The fast moving streams of the valley had not yet cut out flood plains; consequently, the banks of the rivers are as steep as the mountains arising from them.

Fig. 4.11. Looking out and down into a deep, narrow valley

I had noticed many vertical granite outcrops that begin at the base of the mountains and quickly disappear underneath the thick vegetation as the mountain ascends. The drainage system represented by this terrain is dendritic, i.e., branching, treelike in appearance. As I looked at these mountains, I thought of the Appalachian Mountains where I had grown up in West Virginia. The Appalachians are now rather low, rounded mountains. The highest is not much over six thousand feet. They were formed during the Paleozoic and early Mesozoic Eras and were, for millions of years, much higher than they are today. I could not help but think of time and the wearing down and leveling of mountains in one place, the orogenic rise of other systems somewhere else on the earth, a process known as isostatic balance.

About 3 o'clock Sharp lined the people up for census. In addition to this village, people from a village higher in the mountains came down for the census. We believe the inland people may represent the original people that came to the island and were subsequently pushed into the mountains by latter arrivals. This, of course, is only one hypothesis. There is much to be done in New Britain physical anthropology, and I hope to return someday to continue my research.

This is a high village, also built out on a spur and surrounded by a fence. The houses' roofs slope nearly to the ground. At night a low fire is built in the center of the house, and the family sleeps around the fire with only a light cloth over them. The smoke passes through the grass roof during the night. Today, I noticed that some of the blankets used for sleeping appear to have been acquired from traders.

We had another slight quake around 3:30 today that lasted for some thirty seconds. After, the wind came up the mountain from the southeast and blew the clouds away. A *singsing* was begun in the village. When we returned to our house about 10 o'clock it was cold. I will sleep with my jacket plus two wool blankets.

May 22

We left for Lugi, our next village, about 8 am. The weather was clear and sunny. Since Kaiko is perched on a high mountain spur, we had to descend the mountain into a valley, go up another mountain and then down into another narrow valley. The river at the bottom was rather fast, about waist high, with rapids and a rocky bottom. I crossed with the help of our carriers. After that we were having a relatively smooth walk, but when we came around a small hill, suddenly there it was before us: the tallest mountain on our trip!

The trail was steep, slippery and treacherous. We climbed slowly, entering Lugi about 11 am. This village is not surrounded by a stockade. The houses are small and built on the ground. We stayed about an hour while Sharp took census, then went on to the next village located slightly lower than Lugi on the same mountain. When we arrived, it started to rain and continued to rain for hours and got colder. This village did have a stockade. Sharp took a quick census. The fog became denser as evening became night. It was again a two-woolen-blanket night.

May 23

The morning was beautiful and the mountains were spectacular, reminding me of the Swiss Alps, *sans* snow. The next three villages are close together, and Sharp took census in them all. After dinner there was a *singsing* in the village, but we were too tired to attend. Sharp and I talked for a while, and I admitted to him that my hemorrhoids are getting more troublesome and that each step is painful and, on top of that, I am beginning to feel unwell. I had delayed mentioning it as long as I could.

May 24

We were up at 6 a.m. I stayed in bed most of the day. The natives killed a pig in the afternoon and served a great dinner that evening. I am not feeling any better, and it has been decided that I should go back to Galilo. Sharp and I talked a long time that evening about it. I am extremely disappointed, as I am otherwise having a great time and seeing a lot, but obviously if I cannot walk I can not go on.

May 25

We were up at 6 o'clock for breakfast, and Sharp prepared to go on to the next village. I thanked him and said goodbye, then left in the opposite direction with two men he had selected to escort me back to the coast. The route back is more direct as there is no census to collect. The trail was not too bad. Every step I took was painful, though, and in addition, it started to rain. We stopped frequently because it was difficult for me to walk more that a few yards at a time.

About sunset we arrived at the village of Sipa, half way to the coast. I was totally miserable by then, but full of thanks for the two carriers who brought me this far. I paid them, then had a dinner of Bully Beef and rice after which I went to my bed. The rain on the tin roof will help me to sleep.

May 26

Two men from Sipa would escort me today. The trip was not too bad since it was mostly down hill. We arrived in Walo about noon, had lunch and rented an out-rigger canoe. Two other men said they that would take me to Terobe across the bay, a trip of about three and a half hours, afterwhich I would pick up a new crew to continue in the outrigger to Kapuri. We left almost immediately and had rain all the way across the straight to Kapuri. We arrived about midnight and rested for a few hours. Dinner was another tin of Bully Beef and some coconuts.

After the rain stopped the water calmed down, so we set out again and put up a sail that would speed our trip across the bay to Galilo. I sat in the middle of the canoe and held to the sides as we were blown across the bay. It was an exciting ride and I was glad to see the beach in front of Galilo in front of us.

May 27

We landed at dawn. Ward heard of my early return and came down to the beach to see what was wrong. I told him about my hemorrhoids and why I had come back to Galilo. We paid off my crew and had breakfast then I went to bed where I stayed the rest of the day, only getting up for dinner. Val and Edy came by to see me and brought pain pills. That evening, Ward and Ann spent some time with me talking about my trip, but before long I was drowsy and they left me to go to sleep.

May 28

I was up early, had breakfast and went right back to bed. I did not have lunch, but I did manage to get up and have dinner. We had brought a good many medical supplies with us but, unfortunately, nothing for hemorrhoids. We decided that rest and soaking my rear end in a pan of warm water was the best thing for me. That is essentially what I plan to do for the rest of the month.

May 29——30

I've spent most of these days in bed, or close to it. I am feeling better and decide to get back to work on May 31.

Today, May 30, was Ward's birthday, and the Valentines came for dinner and a little birthday party. Uoso concocted a delicious dessert from bananas and the milk and the meat of a young, fresh coconut. During the party he brought it out and presented it to Ward as a surprise. The Valentines remained until about 11 pm. I excused myself early and retired about 9 o'clock.

May 31

I got up and had breakfast, but my aspirations to go back to work today did not materialize. I felt much better, though walking was still painful so I decided to stay on the porch and relax. As it turned out, I spent most of my time in bed.

During the afternoon a man came by and told me that a volcano had erupted in the Talasea region. He did not know if any one was injured. New Britain has many active volcanoes in various regions and eruptions appear to be fairly common. I immediately thought of Sharp and the villages we had passed through, but then I remembered that Talasea was a long way from the middle of New Britain.

On my way to the outhouse this evening, I noticed a scorpion in front of me, apparently headed the same way. I sped up, thinking to step on it, but it also sped up and zipped into the outhouse before I could get to it. I carefully entered and looked around inside the building but did not spot it, so I had to assume it had crawled into the hole. I immediately abandoned my plans to use the outhouse in favor of the bush, and after I alerted Ward and Ann to what I had seen, we unanimously decided to not use the "john" for several days!

I plan to commence my work again tomorrow. I am far behind schedule and now declare my convalescence _over_.

CHAPTER 5

June 1

I was up early, had breakfast and for the first time in a couple of weeks I felt pretty good. After breakfast I sent a note to Val in Rapuri asking if he could muster up several men to come to Galilo for dental impressions. About 9 o'clock five men appeared at our house. I took impressions of each and poured up the casts. That was it for me. I spent most of the day in bed. Ann went clamming with some of the men, and Ward cycled to Moramora. It was a beautiful day, sunny and warm with clear skies. Ann returned about 5: 30 with clams and Ward got back at 6 for a delicious clam dinner.

June 2

We had breakfast and I went back to bed, a habit I was hoping to break soon. Ward and Ann spent the day working in Galilo. I am beginning to wonder if I will ever get better since I am in pain every time I get up and walk a bit. I still have a lot to do and am getting eager to get back to work.

June 3

After breakfast Ward and Ann left for Rago for the day. I am feeling better and hope that I can make it through the whole day. About 9 o'clock several men came to the house from Rapuri to have their dental impressions taken. After lunch I made several more. It was then that I realized I had enough casting material for only about six more impressions. My collection would have to stop at 48 casts. Unfortunately, the impressions were all of males since up to now the women only watched. As it happens, lately I have begun to hear that some of the women in the village want me to make impressions of their teeth. To my disappointment, it is now too late. I'm essentially out of material. I had bought as much dental material in Sydney as I could carry given the weight restrictions and it was impossible to get any more sent to me from Australia.

This evening Ann went to bed early, but Ward and I stayed up till 1 o'clock. We talked about my trip with Sharp and what I had seen. It is a hot night and there is no breeze. Ward said he thinks it is finally summer!

June 4

Ward and Ann went off this morning to collect more data about Lakalai kinship and social organization. About 9 o' clock several men came by for me to make dental impressions. I made five, then explained to the other men that I was out of dental supplies and could not make any more. Later in the afternoon some children came by, and I gave them the little Albastone that was left to mix with water and play with.

Several women from Galilo came to the house so I could make casts of their teeth. I had to tell them, too, that I was out of supplies. They were disappointed and so was I. It would have been a better representation of the Lakalai dentition if there had been female casts.

Ward and Ann came home for lunch, neither one of them feeling well. Ann, in fact, spent the afternoon in bed. Ward stayed on the porch, reading. Before dinner he said he was feeling better, and he and I went for a swim. After dinner we sat on the porch and talked. The evening was hot with a little breeze, and the stars were so bright that you felt you could touch them.

About 9 o'clock Baba came by with three fish for us. He is known as the fisherman of the village. He has "powers" for catching fish that he got from his maternal uncle, and he always seems to get fish when others do not. His powers must be strong for he keeps us well supplied. He wanted to tell us one of his secrets that he used to catch them, but before revealing it he got up and walked around the porch to make sure no one was listening. He came back to the table, sat down and whispered to us, "You tie a possum's tooth to the line above the bait with the tooth crown pointing toward the bait." He asked us not to tell anyone in the village and we said we would not. We talked for a little longer and when Baba left, he said he would bring us more fish tomorrow.

June 5

We had breakfast about 7 o' clock. Ward went to Karapi to talk with the *luluai* and Ann went with the women to the gardens. During the morning I worked out my schedule for the remainder of my stay in Lakalai. I still had much to do, but I figured I could get it done.

During the afternoon I collected ten blood samples. I have had no problem collecting blood samples from the men although no women have volunteered. I had hoped to collect more blood samples for our studies of blood groups and check for the presence of sickle cell anemia. The latter study, regarding the presence or absence of sickle cell anemia in the blood, would be particularly important since this region is a hyper-endemic malaria area and it is known that the sickle cell trait (heterozygous condition) has a selective advantage in malarial regions. The sickle cell is widely distributed throughout Africa and had recently been found among the Veddoid tribes of Central India but, as far as I knew, has not been studied in Melanesia.

I have waited until now to try to take hair samples. Once I became aware that hair—along with fingernails and excrement— is frequently believed to be use in sorcery, in the casting of spells, I realized I would need the full confidence of people to even ask for a sample of their hair. I assured them today that no one but me would ever see, much less touch, the samples, but many were still reluctant. My hair sample will be quite small, I fear.

June 6

I must have been more tired than I thought as I slept until 8:45, the latest I have slept since arriving in New Britain. We went to church, after which we walked home along the beach and watched the children swimming and playing in the ocean. The water here is warm and the beaches are soft, composed of volcanic sands. Both boys and girls learn to swim very young and by the time they are ten or twelve are excellent swimmers (Fig. 5.1).

Fig. 5.1. The swinmmers

The afternoon was hot, but a little breeze blowing in from the ocean made it quite comfortable. The breeze is always welcome since it helps to keep away the insects, particularly mosquitoes. We spent the afternoon writing letters and checking field notes. About 4:30, Val and Edy came by and had dinner with us. They had gone to the egg fields on Saturday with some of the women of Rapuri and watched them descend into the thermal holes to gather bush fowl eggs.

The Valentines left about 8 pm and Baba appeared about 8:30 with four fish. We thanked him and he sat down to talk the rest of the evening. He is an excellent source of information about what is going on in Galilo as well as other villages. He enjoys gossiping and anthropologists enjoy listening.

June 7

Kalua came by after breakfast, and we walked to Rapuri to measure several women. I was walking much better by this time. We used Val's house and measured six women before lunch. They were very pleasant and talked and laughed while I measured them. We ate lunch with the Valentines and had fried taro, a specialty of their cook. The taro is finely graded, then rolled up into tiny balls and fried in pork grease. It is delicious and so far it was the only way I have been able to enjoy this sticky tuber.

We had to wait until 3 o'clock to measure five more women since they were working in the gardens. After we finished, we walked back to Galilo through Vavua so I could inquire if I could measure there tomorrow. I was told that there were ten women to be measured tomorrow. I thanked the *luluai* and told him we would be there at 9 o'clock. Kaluai and I walked back to Galilo.

Not long after I got home a strong wind started to blow from the northeast. It was so strong that we decided not to take our daily swim. After a dinner of delicious *kuka* (crab) we were sitting listening to the wind howl through the tall coconut palm trees when suddenly the hanging lanterns began to sway to-and-thro and the house started to shake. We did not say anything, just looked at one another, and then it was over as quickly as it had started. In a few minutes some men came by and told us that it had been a *liklik guria* (little earthquake) and that there was nothing to worry about. We thanked them.

Ward and I smoked a cigarette to relax before going to bed.

June 8

When Kalua and I arrived in Vavua, the people of the village were already preparing for a *kaikai* (feast) for a man who died a few months ago. It was the man I had met before who had been seriously ill with a stomach problem. He'd had an enlarged abdomen, perhaps cancerous, according to Bill Bell. Many of the villagers, however, thought he died because he disobeyed certain taboos, eating and drinking things he should not have. Because of this, they said, spirits entered his body and his stomach became large, and eventually he died.

We were invited to the *kaikai* that evening, and in the meantime I would measure the women.

Today, on our walk back to Galilo, Kalua told me that he was going to carve a small outrigger canoe for me to take home like one I had watched them make in the forest (Fig. 5.2). I told him I would like that very much.

Fig. 5.2. A dugout canoe under construction in Galilo

People from Rapuri, Ubai, Koimumu, and Galilo attended the singsing. Those from Galilo sat together; others sat with people from their own villages. The meal — roasted pig, taro fried with eggs, roasted chicken — was good and everyone enjoyed themselves. There was a lot of talking and laughter among the people, and I believe much of the talk was about the man who had died. I wondered how different this festival is from an Irish wake? Later, I was told that his widow, who had been in mourning since his death, could now remove the several bands she had been wearing on her legs as sign of mourning if she wished to or could retain them and remove them later—the decision was up to her.

I thanked the *luluai* for inviting me and said good-by to the villagers, then I walked back to Galilo where there was also a *singsing* at which I wanted to play the drum. When I arrived, Kalua handed me a drum and said the people wanted to hear me play. As I began to feel the rhythm and the singing got louder, my drumming got louder and everyone was having a great time. I was beginning to think of myself as the Gene Krupa of the Bismarck Archipelago. It was a long night, one I will long remember.

On the way home, Kalua told me that everyone liked the way I beat the drum and that I would always be welcome at *singsings*.

June 9

Kalua came by and we headed off to Makasili. During the morning I measured ten women. It is interesting that the most difficult measurement in both sexes is sitting height since it is so difficult to get people to sit straight. I suspect this measurement will have a high coefficient of variation!

We had lunch and later I obtained twelve samples of blood. I have been fortunate in having no problem taking blood from either sex. In this village, three little girls—perhaps between six and nine—came up to have blood taken. They presented their fingers, I pricked them and there was no crying. Many people who had gathered around were smiling and talking among themselves, and I believe, after that, I could have taken blood from the whole village. I suspect that many people in the Lakalai villages are anemic since they all have malaria. Many deaths before five years old, in fact, are due to cerebral malaria. After that age, those that do not die become immune to the disease and continue to live with it.

I walked back to Galilo. When Ann returned from the gardens we had a swim. After dinner I went to a *singsing* in a neighboring village. When I arrived it was about dark, and I was given a drum and asked to sit down and play when the dancing started. Soon after the dancing and singing commenced the moon started to rise and tonight it was about half full, amazingly large and clear. The *singsing* lasted for about three hours, everyone had a grand time dancing and singing, and I played the drum the entire time. After it was over I was told that during the next *singsing* I would have to dance as well as play the drums. I accepted the challenge. I must confess I felt completely at ease, mentally relaxed and in another world as I sat in a circle with the other drummers, chewing betel nut and keeping the beat on the drum with the dancers' bare feet as they glided across the ground. This had been a night to remember.

Ward returned about the same time I did. We sat up, shared a large papaya and talked till after midnight. As I retire, it is starting to rain.

June 10

Kalua picked me up after breakfast and we started to bike to Karapi, but about half way there my bike had a flat tire. Kalua rode on to tell them that we would be there tomorrow. I walked my bike back to Galilo and Sege said he would fix the tire. After lunch I worked on organizing my field notes and measurement data sheets for the various villages.

In the afternoon I obtained twelve blood samples and measured four individuals. After that I spent the rest of the afternoon writing letters. After dinner I walked along the beach for a while thinking about the last few months and all the people I had met since leaving Philadelphia. Before too long it started to get dark, so I walked to the village for another *singsing*. As I entered Galilo a young man came running up to me and handed me a drum. I beat it a few times and walked over to the edge of a growing crowd of people and sat down. Before long several men approached me and asked me to move closer and sit beside Bega who was playing a *gita* (guitar). Shortly, there was a group of young girls and boys surrounding us, singing and dancing, and before long one of the men offered me betel nut. I accepted and then I realized, here I am again, playing a drum, chewing betel nut, surrounded by a large group of Lakalai dancing and swaying to the "enchanting rhythm" beat out by Bega and me and punctuated by the sound of bare feet thumping on the ground on an island in the middle of nowhere. There was no doubt in my mind why I was an anthropologist. I stayed at the dance for three hours, by that time my arms and hands becoming too tired to continue drumming. I told my friends I had to leave and said goodnight. I heard them singing and dancing as I walked home, and then I reminded myself what an Australian anthropologist had warned us about in Sydney when he said, "Be careful lads, don't go 'native'." I wondered at the time what he meant by that.

June 11

I measured ten women in Karapi during the morning, and as I worked the women would ask me questions in "talk place" (Lakalai), and Kalua would tell me how to answer the question in Lakalai. As I attempted to answer them in Lakalai they would laugh, and I would pretend to be embarrassed which would bring on more laughter (Fig. 5.3).

Fig. 5.3. Taking "head height" with a Western Reserve head spanner

One woman started to cough when I measured her head height since the measurement required inserting the ends of the head-spanner into the ear openings. This stimulation of the ear will, on occasion, make one cough. When I removed the instrument she began to laugh and said the device made her cough. My Lakalai isn't good enough to explain the neurological explanation for this. The vagus nerve has the most extensive distribution of all the cranial nerves. One small branch passes to the outer side of the eardrum and when stimulated can cause a tickling or coughing reflex. The Romans, during their huge feasts, would sometimes stimulate that branch of the vagus so as to vomit so they could keep eating. Kalua said something to her, I don't know what, and everyone laughed.

I only had about another month in Lakalai, and I wanted to do as much as I could before I left. I was beginning to have mixed emotions about leaving the Lakalai whom I had grown so fond of in the months I had been here. I knew that I would probably never return.

After dinner I walked over to Kalua's house and talked with him and his wife for a couple of hours. I returned home to find Baba talking with Ward. He had brought a couple of fish and stayed to talk for the evening. They were still talking when I left to write up my notes and get ready for bed.

June 12

I spent much of the day getting my gear ready for a two week trip that would include stopping at the plantation owned by Maynard and a visit to the Catholic Mission near Valoka, about 20 miles northwest of Galilo. I had discussed this with Maynard and Father Berger when they visited us about a month ago, as setting up a temporary base with them was the only way I could work at the Lakalai villages so far to the northwest. I would spend several nights at Maynard's and then move on to the mission from where I could cycle to the several Lakalai villages to the west.

After lunch, I checked the blood samples I'd taken 48 hours ago for sickle cell anemia. They were negative for the trait. So far, all blood samples have been negative, which is somewhat surprising since this is a malaria area.

Later in the afternoon Sege and I walked to Kulau's house and sat with him and chewed betel and talked for a couple of hours. I wondered, Is this *'going native?'* We discussed the effects of betel. I have chewed betel several times and never felt "high," although the Valentines have told me that they received a little "boost" from it. I have noted feeling a slight irritation on my cheeks, however, and my teeth turn dark red from its juices.

Sege and I returned to the house where he helped Uoso get dinner. Tavili came by later to see Ward and Ann. They talked about some of the tricks and games the Lakalai play. He said they have many string tricks, such as "cat's cradle," as well as riddles they ask one another. I mentioned that I had seen children making various designs of "cat's cradle" that were quite involved. Ward and Ann said they had watched the same thing. It was an interesting evening and Tavili left about 11 o'clock.

June 13

When Kalua arrived, we cycled to the village of Valoka. Here we had lunch with Father Munslinger, who was stationed in the interior of the island, and Father Berger, from the Catholic Mission near Valoka where I was planning to stay for week. Father Munslinger is a rather nervous individual about 50 years old, who agrees with everything one says to him. He had been in New Britain since before World War II and experienced many problems with the Japanese during the war. He is very pleasant and speaks English well. I enjoy his company, but can see that he is ready to return to the mountains.

After lunch we cycled to Maynard's Plantation at Matavulu. Father Munslinger stayed for half an hour and excused himself saying he had to return to the Mission and pack since he planned to leave tomorrow. We said our farewells, shook hands, and he left. Father Berger and I talked about what I wanted to do in this area of Lakalai. Before he left he said he would talk with the local villagers and explain to them what I wanted to do. He was sure that I would be able to work with them.

Maynard and I had dinner and then sat on the porch for most of the evening. His house was perched on a low bluff and had a magnificent view of the Pacific Ocean. It was a beautiful evening we had a couple of cocktails. By now, I had become accustomed to drinking *wisiki wantaim wara*, usually rum with water, no ice, and after we had a couple we were pleased to discover that we had solved most of the problems of the world.

Maynard told me he had arrived shortly after WW II and started his plantation which today has approximately 650 square acres that he leases from the Australian government for a period of 99 years. He produces mainly copra and cocoa and is very content with his life. He said he left Australia because of what he considered to be deplorable conditions after the war and would never return. Maynard is a very interesting person. He likes his independence and freedom and will probably spend the rest of his life in New Britain. He told me that there were over 60, 000 acres of fertile land between the mountains and the coast in this area of New Britain that could be cultivated if people would move in and start clearing the land. He says the land is excellent for coconut cultivation as well as many other profitable products. It doesn't matter to him who does it, local people or outsiders, only that people take advantage of this fertile land.

It was well after midnight when we decided to retire. It has been a long day, and I know I will go to sleep as soon as I slip under my mosquito net.

June 14

I arose before 6:30, but Maynard was already up and getting breakfast. This morning we had bacon, eggs, papaya and tea. After breakfast Kalua came by and we cycled to Kovalakese, about 2 miles from Maynards, and the most western of the Lakalai villages (Map 3). It is a rather large Catholic village with about 150 inhabitants. The *luluai* met us in the center of the village. He had already arranged for ten people, nine men and one woman, for me to measure. I thanked him, and as I started to work I soon noticed that he had quietly left. I never saw him again.

About the time I measured the last individual it started to rain. I had been working under a large tree in the middle of the village, but moved to a spot under a large house built on rather high poles in order to begin collecting blood samples; a rather more comfortable location for my "blood letting."

Today, I obtained blood from the nine men that I had measured and a woman "volunteered" after a little joking and some good-natured prodding by the other women.

After lunch, the men wanted to talk about World War II, the Japanese, and the Americans who replaced them. They liked the Americans, but were also happy when they left because, when they did, they left all of their equipment behind. They also wanted to know how far away America was from their island? The Lakalai were aware that there were many countries in the world, but it was understandably difficult for them to grasp geographical distances.

During their lifetime several older individuals living here today had seen the missionaries arrive, then years later the Japanese occupiers, who were in turn replaced by the American soldiers and then the Australians who took over government of the island. This had been a time of many cultural and political upheavals and I am afraid the effects of them are not over yet.

June 15

Kalua and I cycled to the village of Kasai, only about a mile from Maynard's Plantation. During the morning I measured ten men and then took blood samples. We had lunch of sweet potatoes, salad and fresh coconut milk to drink. The sweet potatoes were boiled in coconut milk that enhanced their flavor. The people of Kasia were very friendly and cooperative and, if I had had the time and supplies, I could have measured and taken blood samples from everyone in the village.

After lunch we sat and talked for a while and, as usual, there were questions as to whether I had been here during the war and if more Americans were coming when we left. I said I didn't know if more Americans were coming after we left, but perhaps some of our people would be back someday.

June 16

A crowing rooster woke me up at dawn. I looked for a gun, but found none, so the rooster was safe. Kalua and I cycled to Galeoale, a village on the beach about a mile from here where I measured ten men in the morning and took blood samples. When I returned to Maynard's he had just returned from the fields and was sweaty and thirsty. After a couple of beers I took another shower and wrote up my field notes. Tomorrow I will go to the Catholic Mission.

June 17

After breakfast Maynard drove me to Valoka. On the way we met Kalua who had kept my instruments with him at Galeoale where he had stayed the night with sib members. I asked Maynard to stop and pick up my instruments and Kalua. Maynard said we could pick up my instruments, but surprised me by saying he would not permit Kalua to ride with us. I was polite, but asked Maynard to stop and told him I would walk to Valoka with Kalua. When we arrived at the Mission, Maynard was waiting. He said he was sorry, but it was his policy not to let the locals ride with him. I said I understood, even though I was not sympathetic to his point of view. He stayed and had a cup of coffee with Father Berger and me, then left.

After dinner Father Berger told me about his own situation during the Japanese occupation. He'd had several bad experiences but would never explain more than to say that they were "bad times." He also told me that he felt he was being blamed for some of the current problems among the Lakalai but it was his opinion that Boas was at the center of the native problems while Lima was trying to help the natives. He was not very specific about what the problems were, and I wondered if he was referring to the Cargo Cult.

We talked for a while, and then he asked me if I would like a drink. I accepted, and he called to one of the Sisters who brought a bottle of rum and one glass and proceeded to pour me a drink. I then asked Father Berger if he would join me, and he accepted, asking the Sister to bring another glass and pour him a drink. When she left she left the bottle and we sat there during the remainder of the evening, talking and drinking.

It was a delightful evening, the moon low in the sky, the stars shining through the Milky Way, the Southern Cross outlined in the sky. We talked until after midnight and then retired.

June 18

The next morning I had a slight headache and didn't get up until 9 am. The Sisters made breakfast for me and packed a lunch since I would not be back to the Mission before dinner. Kalua and I left for Porapora, a village about a mile from the Mission. I measured eight men before it started to rain and we were invited into one of the houses to complete our work. We finished about

3 pm and returned home, and when we returned found that the Sisters had cakes, a type of tapioca, and coconut milk for us to drink. Later, I sat down to write up my field notes, and I wrote that I could see no physical differences between these Lakalai and those to the east. They look to be the same, and I am certain my measurements will support these observations.

June 19

The Sisters had another wonderful breakfast for me as well as a packed lunch. We were off to Vuvusi, another beach village to the west of the Mission. I measured eight men during the morning and about 10 o'clock a Chinese fellow who lived in the village brought me tea and cookies. I thought, *Boy, this is tough work.*

After lunch I cycled to Gavaiva where I measured eight men and obtained blood samples. I returned to Valoka about 4:30, showered and had dinner with Father Berger. We sat on the porch after eating and Father Berger asked me if I would like a drink. I accepted, and we went through the same elaborate ritual as before.

June 20

Today was a special Sunday service, for today the Catholic Church celebrated Corpus Christi (Body of Christ) in honor of the Eucharist. People came from many surrounding villages to attend the celebration (Fig. 5.4).

Fig. 5.4. People entering the church in Valoka

The procession started with four altar boys dressed in blue and white robes; three of them carried golden painted wooden crosses while the fourth swung a metal bowl containing incense. Father Berger was very impressive, dressed in beautiful red and yellow robes that flowed to the ground as he preached his sermon in Pidgin. There was much singing, bowing, genuflecting, and bell ringing during the two-hour service. Everyone enjoyed the festivities and everyone participated.

At the end of the service people lined up for Holy Communion after which they marched out of the church and waited for Father Berger. He came out under a large yellow canopy embossed with several brown crosses and carried by four young boys. Altar boys walked in front of him carrying lit candles. As the procession snaked its way around the yard a bell would ring periodically signaling everyone to stop, knell and chant before moving on in this seemingly endless, serpentine parade. The procession lasted about an hour and then gradually started to break up as the people began to leave to return to their villages.

After the ceremony I cycled to Sharp's to spend the rest of the day. We talked about our trip to the mountains, and he expressed his regret that I had to return early. Mavis cooked a wonderful lunch that was topped off with a chocolate cake. By the time I had cycled back to the Mission it was time for dinner. After dinner Father Berger and I sat on the porch, and in a while one of the Sisters brought a bottle of rum and a glass. Again, I accepted the drink, and then Father Berger said, "Since you are having a drink I shall join you." The Sister left the bottle and brought another glass.

We sat on the porch talking well into the night. He told me some of his experiences with the people when he came to the island in 1937, and talked about the New Britain people. He described some as not very tall in stature with light skin and light brown eyes. He believed the girls in Lakalai, as elsewhere in the Tropics, do not reach puberty until they were about 17 years old; but as far as I know there are no scientific data to support this assumption.

Tonight Father Berger was very talkative. He said he believed that Boas had cooperated with the Japanese when they occupied the island. Although I had not expressed any interest in this problem, he said he wanted me to know that there was no "Cargo Cult" in Lakalai, and if there was, it was not among the Catholic villages. He also told me that the intertribal problems are not due to religious differences, but to the poor administration policies and the management of native affairs. He said that all he has ever wanted for the natives was for them to understand the idea that the family was the most important part of their culture and for their tribal solidarity. He said he knew he has been accused of many things by the government but he was innocent. It was a most interesting evening.

June 21

I cycled to Kololo, a village a little east of the Mission and by late afternoon had measured ten men, stopping for tea sent by the sisters, and lunch.

June 22

I was up early this morning and off to Kololo again. When tea arrived from the Sisters I found myself wishing that their Mission had been built closer to Galilo!

The people in the village were very friendly and, when I finished measuring the last person, the *luluai* of the village asked me to stay for a while. I sat down in the middle of the village on a bench and he came up and presented me with bananas and papaya. We talked for a while and then I left for Voloka where I measured 12 men during the afternoon.

June 23

The Sisters had breakfast for me before Kalua and I cycled to Gavuvu, an hour away, to measure and collect blood. I wrapped up all the blood samples that I had taken over the last few days and stored in the Mission refrigerator, for this would be my last day there. Before I left, I thanked the Sisters for all their help and kindness.

Ward greeted us on our return. Ann was clamming with some of the villagers and clams were on the menu for tonight's dinner.

June 24

Today was a rest day for me. About mid-morning Kalua came by and had me follow him into the bush to see a large fish he was carving from a log for a *singsing* that would take place in a few days. The carving was being done in the bush so no women could see it before the *singsing*. In the "old days," he said, if a woman accidentally saw men carving an object for a *singsing* she would have been killed. Even today, a woman would be beaten if she were discovered watching the carving ceremony.

The man who requested the carving is obliged to kill a pig and give a feast or the fish will break into many pieces. It would take about six hours for the fish to be carved, and then it would be painted. Kalua is a master carver and let me watch him work (Fig. 5.5).

Fig. 5.5. Kalua carving a fish with his helper

June 25

The morning was beautiful, and some women came by our house to be measured which I took care of before going to Rapuri to visit Val.

After my visit I watched the preparation in Galilo for the forth-coming *singsing* to be held in honor of a man killed during World War II. This is one of the most important ceremonies held by the Lakalai. Several pigs would be killed for the feast and there would be much crying and mourning.

The dancing this evening consisted of two groups of men performing. One group formed a straight line while the other group lined up at a right angle. The two ends moved away from each other, stamping their feet and waving branches to and fro while singing.

June 26

I have been having trouble again with hemorrhoids and am beginning to consider leaving Galilo for home on July 7th rather than July 27th. I believe I can finish my study in another week. I already have measurements of a large population of the Lakalai as well as blood samples, fingerprints, and dental casts—enough data for my Ph. D. dissertation, I hope!

There is more than the degree involved in this decision, of course. I have become very close to the Lakalai, a people that I know, respect and love, people such as Sege, Uoso, and Kalua whom I will never forget, as well as the many, many people who willingly have helped me.

Today I told Kalua that I had decided to start back to America two weeks early. He asked me, "Why don't you stay in Galilo? I can get you a wife here. There are many girls to choose from." I thanked him, but told him that I already had a wife in America and that she was waiting for me to return. He understood but seemed downcast and said he might "go bush" when I left. He will not, of course. Without his help, I don't believe I could have collected as much data as I did. Whenever my attempts to explain to the people what I wanted to do failed, Kalua had been there to help clarify everything and pave the way. I hesitate to guess how many times I had pricked Kalua's finger demonstrating the blood collection routine.

June 27

Today I got my "traveling" clothes out and had a look at them. My suit has many mildew spots on it as did my shirts. I removed as many of them with lighter fluid as I could, but quite a few remain. They will have to do until I get to Rabaul or Port Moresby where I can buy some new clothes.

I spent the afternoon working on a few statistical analyses of the measurements. Later in the afternoon a group of children came by and I stopped working to talk with them. I amused them with a few hand tricks and they taught me a few more Lakalai words. They stayed for about an hour and, as they left, Ann came back from the gardens.

Ward had not returned yet from Moramora to see if Bell had arrived from Talasea with mail, but Val and Edy came by, and Ann and I sat with them and talked all evening. Val was collecting some interesting information from Lima concerning the Cargo Cult, the topic that had worried Father Berger so much.

It is a lovely, warm night, so after they left I lit a cigarette and walked down to the beach. I sat for a long time looking at the stars, seeing how many constellations I could find before going back to the house and to bed.

June 28

After breakfast Kalua and I cycled to Koimumu, Karapi, and Vauva where we collected a total of 40 blood samples. I had been to these villages several times, measuring, getting finger prints, and making dental impressions, but purposely waited to the last before starting to collect blood samples in these other villages until the people had the opportunity to get to know me.

We spent most of the morning cycling and collecting blood

samples and then had lunch in Rapuri with Lima and the Valentines under a large palm tree. After lunch Lima agreed for me to take a sample of his blood. I did, thanked him, and then told him that he was now part of my study of the Lakalai people, which seemed to please him very much.

On our way back to Galilo, I asked Kalua about the different hair colors present among the Lakalai. He told me that many people use a Peacock Black Hair Dye made in Japan. They also use peroxide to dye their hair blond and reddish orange. Both of these colors are common among the Lakalai and people buy these from the Chinese traders from Talasea. I asked him why he did not use them and he told me that when he was younger he had, but now he liked his hair *olosem blak* (naturally black).

June 29

After lunch Kalua and I walked into Galilo where I measured ten women. That brought my female sample to 75. This is my last day for anthropometry in Lakalai.

After dinner I talked for a while with Sege and Uoso about going back home. They knew I was getting eager to leave and wanted to see my wife. I told them it is true, I want to go home, but that I will be sorry to leave since I like the Lakalai people very much.

June 30

Several men were waiting for me when I arrived in Karapi and Makasili to collect blood samples. I thanked the *luluai* in each village and said good-bye to the people who knew that I was leaving soon. I told Kalua how pleased I was with all the information we had collected. We were both happy and sorry at the same time.

Bill Bell was there with our mail when I returned. He stayed for about an hour and we made arrangements for me to go back to Talasea with him on Sunday. I will stay with him for about a week before catching a plane to Rabaul. During this time he said I would be able to collect additional blood smears for my sickle cell study from the people living around Talasea.

We talked a lot that night about the expedition and how we had looked forward to it for so many months, and how it would soon be over for Ward and me. Ann and the Valentines are staying for several more months. It will be difficult to sleep tonight, with so many things on my mind. I even find myself wondering where "my" spider is and how he is getting along.

CHAPTER SIX

July 1

I t was a beautiful morning, and I took a short swim before breakfast. After breakfast, Kalua and I walked into the center of Galilo and collected 20 blood samples. That brings the sample to 195. This is pretty good, since I had brought only 200 flasks with me from Melbourne. These will now be sent back to the Commonwealth Serum Laboratory in Melbourne for analysis.

There was the customary joking and "wise cracking" among the villagers, one person suggesting a friend take his place, but giving in when this friend or another started to tease him. Today one of the men disappeared when it was his turn to "donate," reappearing from behind a house when he was called saying he had gone to *pispis* (urinate) and was coming back. Everyone laughed, and when he presented his finger to me he said very softly "*katim isi*" ("cut easy"). Of course, when the people heard this they started laughing again. I cut easy and all was well.

During the afternoon I worked on my field notes, checked my

blood smears, and started wrapping dental casts in toilet paper for the trip back home. Toilet paper makes a great packing material for breakables. I will buy a large suitcase in Talasea to carry the casts in for the trip home.

We had our usual swim in the waters of the warm Pacific. I have thought more than once over the last couple of days how much I am going to miss these swims. After dinner I visited Kalua and took him two *laplaps* as a gift. He thanked me and we sat down and talked. He told me he was bringing me a chicken for my last night's dinner in Galilo. He told me how sorry he was that I was leaving, and I told him I would miss him and would never forget our good times together. Tomorrow, he said, he is carving a crocodile for a *singsing* that was being held in Moramora. I told him I would come by to watch him.

During the evening several people came by to say good-by. It was a pleasant evening, but one of deep emotions.

It is a beautiful night, the stars cover the sky like a blanket of diamonds, and a gentle breeze is blowing from the southeast.

July 2

After breakfast Ann and I walked around Galilo taking pictures and talking with the people. As usual, there were many children walking along with us, all talking at the same time. By now, Ann speaks Lakalai very well and it makes them happy to hear her do it.

After lunch I walked to the jungle outside Galilo where Kalua and another man were carving four wooden lizards for the upcoming *singsing*. It rained the rest of the afternoon, but it was a warm rain.

After dinner I walked to Kalua's house and spent the evening with him and his wife. I took a can of cashews and a few cans of juice, and we had a *"liklik kaikai"* (little dinner). His wife told me that everyone in Galilo would miss me, and I told her that I would miss the people of Galilo. We talked to about 11 o'clock and then I walked home. I admit there were a few tears shed tonight.

July 3

After breakfast I examined 25 blood smear slides for the sickle cell trait. I now have 136 specimens, all negative, and will increase the sample in Talasea next week.

About mid-morning Kalua came by and said he wanted to help me pack. He laughed at the idea of wrapping the dental casts in toilet paper.

We'd finished packing by the time the Valentines arrived for lunch. This was a special lunch since Kalua had brought a chicken his wife had cooked, and Sege and Uoso had prepared bananas cooked in coconut milk, sweet potatoes, and taro. We all had lunch together, five anthropologists, Kalua, Sege, and Uoso.

After lunch, Ward and I put on *laplaps* and Ann took a picture of us standing beside our house (Fig. 5.6).

Fig. 5.6. Ward and Daris in laplaps

A lot of people came around to see us, and most of the women giggled.

We were invited to the Valentines for dinner in Rapuri. The Valentine's cook, Laili, had prepared a large fillet of a corral fish in coconut milk and several boiled lobsters. Of course, there were sweet potatoes and taro, and the Valentines had saved a bottle of wine for the occasion. It was a delightful evening and we talked till after 11 o'clock. On our walk back to Galilo we passed by a tree that was famous for the great number of fireflies that swarmed around it at certain times of the year. Tonight it was spectacular—there must have been hundreds flitting around the tree.

We retired as soon as we got back to our house. It has been a long, busy day and tomorrow is the Fourth of July, the day I am leaving Galilo. It has started to rain again and I am glad for it will help me to get to sleep.

July 4

This morning I had my last swim in this part of the Pacific Ocean. After Ann and Ward got up, we had breakfast and I finished a few items of packing. About 10 o'clock a large out-rigger canoe arrived that took my gear to Moramora. I was to meet Bell there and go in his boat to Talasea.

Ann and Ward were going to cycle with me to Moramora, and as we were about to leave several people appeared, to say good-by. One of the men stepped up to me and when we shook hands he smiled and turned my wrist over and felt for my palmaris longus. I almost cried.

Kalua was among the people saying goodbye and by then we both had tears in our eyes. We shook hands and I gave him a hug. We said good-by and he quickly walked away. Most likely we will never see each other again.

When we arrived at Moramora, Bell was waiting and told me we would depart tomorrow. We had a bite to eat, after which I said good-by to Ann and Ward who were going back to Galilo. The good-byes were not getting any easier.

I sat talking the rest of the afternoon with Bell and Sharp. Sharp mentioned that he had some firecrackers he had bought for the coming *singsing* in Galilo, and since it was the Fourth of July he suggested setting off a package or two in celebration. "Great," I said, and we fired off two packages of firecrackers, one at a time. After the celebration we had dinner and talked till about 11 o'clock.

July 5

We were up at 6:30, ate breakfast and went out to the boat. The engine did not start! The batteries were dead, and charging would take several hours. Bell said we had better delay our departure until the next day to give him time to make sure everything is OK with the boat.

July 6

The next morning Bell said he thought everything was fine with the boat, which I hoped was the case. He started the engine and we left for Talasea, a trip of about five and a half hours. We traveled without incident as the sea was calm and the weather beautiful. The boat is named Alia and is 40 feet long. When we docked in Talasea we were met by Michael Foley who invited me to stay at his place for the night after which I could move into the McMeekin house. The McMeekins have a plantation near Talasea and are going on vacation tomorrow to Rabaul.

I sat in Foley's office most of the day, though I did take a walk around Talasea and bought a small suitcase in a Chinese Trade store for carrying my dental casts. I would carry it with me on the plane as I did not want any of them to break if I could possibly help it. About 5 o'clock Foley and I went to Toong Pins, a local pub, where we had a few beers before going on to Bell's house for dinner, then to Foley's for a going away party he was throwing for the McMeekins. Several people from several of the local plantations were there and everyone had a good time.

July 7

After breakfast we went to the hospital where I spent most of the day collecting more blood samples. Bell said he probably can get me about 100 blood samples from natives living around the Talasea area, and made one of his laboratories available to me. He has been extremely cooperative and helpful during my stay in Talasea. In addition, we are now having a tea break at 10 am and 3 pm, a most delightful habit.

July 8 and 9

These two days were spent much as the 7th. It is a pretty rigid routine, but one that is probably good for me since, when I return to the States, it is back to graduate school for me and that is a rigid routine, if I remember correctly!

July 10

Today is Saturday, which means that the public portion of the
hospital closes at noon. I went home then and read until Bell
came for me about 3:30 to go to Toong Pins where most of the
male Australian population of Talasea gathers each Saturday afternoon for beers, darts, and gossip. Tomorrow we are going to
Garua Island with the Foleys.

July 11

Bell and I walked to the wharf where Foley was waiting for us on
his boat. We sailed to Garua Island which has the largest privately owned plantation in the territory. We met the manager and
his wife, Mr. and Mrs. McMillan, who live in a large, beautiful
house in the middle of the plantation. They were very pleasant
and during lunch we talked about the products of the plantation.
They produce about 90 tons of copra a month. It takes a coconut
palm tree about eight years to mature, that is, before the coconuts can be used for copra. They also plant and grow cacao trees
between the rows of coconut palm trees. It is from the beans that
grow on these trees that cocoa and chocolate are made.

After lunch we took a ride up a beautiful bay for several miles.
It was a lovely trip, the water calm and the bay lined along the
shore with a variety of trees. One in particular drew my attention. Foley told me it is native to Australia and is known as Australian pine. It is called *yal* in Pidgin, he said, a casuarina tree so
named because its leaves are similar to the cassowary's feathers.

We returned to Talasea about 6:30. I had dinner with Foley and
then went home. It has been a lovely day and quite enjoyable. As
I go to sleep I will be thinking about going home.

I worked everyday in my lab at the hospital. The building sits about 30 feet from the beach and is shaded by coconut palms through which run narrow paths lined with a variety of plants and flowers. The casuarina tree is interspersed between the palms. A more beautiful place to work would be hard to find.

Bell had one of his assistants help me, a young man from the village of Gule in Lakalai. We work well together and the time passes quickly. I continue to add to my sickle cell blood smears. He brought in an additional 31 people from neighboring villages which gives me a sample size of 152 for my study. Bell has been one of the nicest and most considerate people here; in fact, everyone has been kind and considerate.

July 18

I was up early and had breakfast with Foley as my plane, a Quantus seaplane, was to leave mid morning for Rabaul. As we walked down to the wharf, I thanked him for all his kindness and hospitality. We embraced, said good-by, and then I boarded the plane. As the plane took off and turned east it flew over the Hoskins Peninsula, and I looked down on the coast where I had spent the last five months and thought of Kalua, Sege, Uoso and all the Lakalai people I had gotten to know and will probably never see again.

The plane arrived in Rabaul where I will spend several days waiting for my flight to Port Moresby, New Guinea. The time now seems to pass slowly.

Finally, the day arrived for me to leave. I flew to Port Moresby where I would spend another day or two waiting for the next leg of the journey home. The trip from Port Moresby was Quantas again, across the Coral Sea and landing at Cairns on the northeast coast of Australia. Most planes coming from New Guinea stop at Cairns to pass through customs. This proved to be an interesting experience for me.

I had collected several mudskippers in a mangrove swamp and preserved them in a jar of ten per cent formaldehyde. I had more or less forgotten about them, but as I approached Customs I remembered that the jar was in my suitcase because a large sign embossed in large black letters above the Custom Officer's desk read:

No Plants or Animals May Be Brought Into Australia From The Islands, No Exceptions.

It was too late to do anything. I opened my suitcases. In one were all my dental casts. I unwrapped one and told the Officer what it was and where I had been for the last several months. "OK," he said. As he was looking through the other suitcase he spotted the glass jar containing the mudskippers. He took it out, peered at it and said, "What are these?"

I told him they were mudskippers, fish from New Britain.

He asked me if I had read the sign above his desk. I replied that I had, but that these fish were not alive and in fact had been preserved as scientific specimens in a ten per cent formaldehyde solution. "Nothing," I announced confidently, "could live in that." He told me, nevertheless, that they could not be brought into Australia. I protested politely and suggested that if he would just undo the cap he would see immediately that this was a solution that would completely sanitize the fish.

The officer apparently was not familiar with formaldehyde so he unscrewed the cap from the jar, put his nose just above the opening and took a rather deep breath of it. As he inhaled the powerful aroma, his pupils dilated and he started to cough and choke. He thrust the jar back to me and told me to get the hell out of there! I thanked him and went merrily on my way.

I re-boarded the plane and flew to Sydney where I would stay two more days before catching my flight to Hawaii.

Leaving Sydney, we had been in the air for several hours when I noticed that Robert Newton, the well-known actor, was sitting across the aisle from me. I had seen him in many movies and was sure it was he. At one point he asked me if he could borrow a pen since his was out of ink. I said, "Of course." He wrote a few cards and handed my pen back. We talked now and then for the remainder of the flight to Hawaii.

We arrived in Hawaii on July 29th.

On July 30, 1954, our plane from Hawaii landed in the United States. I boarded another plane in San Francisco that bought me back to Pittsburgh where I met my wife and my parents.

I have been gone for six months. It is nice to be home again.

Epilogue

When I heard that Daris Swindler was preparing a book based on his New Britain diary, I became curious because I knew that in 1954 he had stayed in the very same area where I had done fieldwork in 1996 and 1997. I asked Daris if I could read what he had written, and he agreed. His memoirs not only proved to be fascinating in themselves, they also reminded me of my own experiences.

Anthropologists, however, are by no means the only ones for whom the past can be brought back to life: on one occasion not long after my arrival in April 1996, I had hardly started to introduce myself and my intentions to a group of villagers when they interrupted me by commenting that 'Vali did that too', referring to Charles Valentine, another participant in the anthropological expedition in 1954. If Daris still remembers his stay in New Britain, the same can also be said for his hosts and their descendants.

Unlike in 1954, it is now quite common for anthropologists to be told about colleagues who have previously worked in the same area or with the same ethnic group. Although such references have not yet received the attention they deserve, they are of interest because, in talking about anthropological predecessors, people may in fact be expressing what they think about their past or about culture change in general.

In 1954, when Daris and the other expedition members went to New Britain, anthropologists had been trained to keep their publications free from any consideration of their own personalities or of their personal reactions to their particular research situations. They could thus only write about their individual hopes and anxieties in the context of their field notes, such as this diary.

Today, anthropologists no longer work the way they did in 1954, but the world has changed as well for the people they visit. The Nakanai of New Britain, for example, are now also citizens of Papua New Guinea, an independent country since 1975. They increasingly wear Western clothes and live in permanent houses, they send their children to school, they keenly listen to the radio and many of them regularly read the newspapers. Thus cash crops, stores and 'public motor vehicles' have become ever more important as essential parts of a growing money economy.

At the same time, however, the majority of the population still plants taro and yams, fishes, collects bush-fowl eggs and processes sago. Moreover, Daris's descriptions of the Nakanai settlement pattern (p. 8) and of their social organization (pp. 9–10) are, by and large, as accurate today as they were half a century ago. When I was living in Koimumu, one of the villages visited by Daris (p. 76), I was particularly struck by the fact that people seemed to be able to delineate the exact genealogical position of almost anyone they met or talked about.

Throughout Daris's diary, one receives an impression of how much he enjoyed the hospitality and generosity with which he was welcomed. Indeed, the Nakanai seemed to do their best to make his work as easy as possible, although, from their own perspective, it must have been quite a challenge to make sense of someone wishing to measure them or take impressions of their teeth, let alone collect blood samples.

Yet, I believe that the Nakanai did appreciate the fact that Daris was willing to chew betel nut, to join them when they went fishing or hunting and to play the drum at festive gatherings. 'As I began to feel the rhythm and the singing got louder', Daris writes in describing one such occasion, 'my drumming got louder and everyone was having a great time. I was beginning to think of myself as the Gene Krupa of the Bismarck Archipelago' (p. 133). Here I detect a sense of humour which, slightly reserved and self-ironical though it may be, does not appear to be too different from that I encountered during many conversations in Koimumu and the surrounding villages.

The very publication of his diary can be taken to show that his stay in New Britain still matters to Daris; indeed, an adventure that he 'would never forget' (p. 19). As it turned out he did not go back. This deprived him of a chance to see things changing and not changing at the same time, but it also saved him from sharing an experience common to many anthropologists: even having re-visited their field sites, they often don't stay home for too long before starting to think about returning yet again.

Holger Jebens

Frankfurt, 2007

Bibliography

Amadio, N. (1993). *Pacifica. Myth, Magic and Traditional Wisdom from the South Sea Islands.* Sydney : Angus & Robertson Publication.

Chowning, A. (1958). *Lakalai Society.* Ph. D. Dissertation, University of Pennsylvania. Available from Ann Arbor: University Microfilms International.

Chowning, A. (1965-66). *Lakalai Kinship.* Anthropological Forum **1**: (Nos. 3-4). pp. 476-501.

Chowning, A. (1966). *Lakalai Revisited.* Expedition: **9**: 2-15.

Chowning, A. (1977). *An Introduction to the Peoples and Cultures of Melanesia.* Menlo Park: Cummings Publishing Co. Inc.

Chowning, A. (1983). Inspiration and convention in Lakalai Paintings. In: Mead, S. M. and B. Kernot (eds.), *Arts and Artists of Oceania.* Palmerston North: Dunmore Press and Mill Valley, CA. Ethnographic Arts, pp.91-104.

Chowning, A. (1989). Sex, Shit, and Shame: Changing Gender Relations Among the Lakalai. In: Marshall, M. and J. L. Caughey (eds.), *Culture, Kin, and Cognition in Oceania: Essays in Honor of Ward H. Goodenough.* Washington, DC. American Anthropological Association, pp. 17-32.

Chowning, A. (1991). Lakalai. In: Hays, T. E. (ed.). *Encyclopedia of World Cultures,* Vol. 2: *Oceania.* Boston: pp.139-143.

Chowning, A. (1995). Future for the Lakalai. Correspondence. *Anthropology Newsletter*, **36**: 2.

Chowning, A. and W. H. Goodenough (1965-1966). Lakalai Political Organization. *Anthropological Forum*, 1: (Nos. 3-4), pp. 412-473.

Goodenough, W. H. (1954). The Pageant of Death in Nakanai: A Report of the 1954 Expedition to New Britain. *Bulletin University Museum*, **19**: 19-43.

Goodenough, W. H. (1956). Some Observations on the Nakanai. *Papua and New Guinea Scientific Society Annual Report and Proceedings*, 1954.

Goodenough W. H. (1962). Kindred and Hamlet in Lakalai, New Britain. *Ethnology*, **1**: 5-12.

Goodenough, W. H. (1971). *Culture, Language and Society.* Massachusetts: Addison-Wesley Publishing Co.

Goodenough, Ward H. (1961). Migrations implied by relationships of New Britain dialects to central Pacific languages. *Journal of the Polynesian Society* **72**:112-126.

Jebens, H. (2003). Starting with the Law of the *tumbuan*: Masked Dances in West New Britain (Papua New Guinea) as an appropriation of one's own Cultural Self. *Anthropos* **98**: 115-126.

Jebens, H. (2004). Talking about Cargo Cults in Koimumu (West New Britain Province, Papua New Guinea). In: *Cargo, Cult, and Culture Critique*, (ed.), Jebens, H. Honolulu: University of Hawai'i Press, pp. 157-169.

Jebens, H. (2004). 'Vali Did That Too': On Western and
Indigenous Cargo Discourses in West New Britain (Papua
New Guinea). *Anthropological Forum*, **14**: 117-139.

Jones, D. N., R.W.R. J. Dekker, and C. S. Roselaar (1995).
The Megapodes. Oxford: Oxford University Press.

Lewis, A.B. (1951). *The Melanesians: People of the South Pacific*.
Chicago: Chicago Natural History Museum.

Luhr, J.F., Editor-In-Chief. (2003). *Earth*. New York:
DK Publishing, Inc.

Mavalwala, J., D. R. Swindler and E. E, Hunt, Jr. (1963).
Fingerprints of the West Nakanai of New Britain. *Am. J.
Phys. Anthrop.*, 21: 335-340.

Murphy, J.J. (1954), *The Book of Pidgin English*. Brisbane:
W. R. Smith & Paterson Pty. Ltd.

Nile, R. and C. Clerk (1996). *Cultural Atlas of Australia, New
Zealand and the South Pacific*. Oxfordshire: Andromeda
Oxford Limited.

Oliver, D. L. (1961). *The Pacific Islands Revised Edition*.
New York: The American Museum of Natural History.

Osborn, F. (1944). *The Pacific World its vast distances, its lands
and the life upon them, and its people*. New York: W. W.
Norton & Co., Inc.

Robson, R. W. (1945). *The Pacific Islands Handbook*. New York:
The Macmillan Company.

Simmons, R. T., J. J. Graydon, N. M. Semple and D. R. Swindler (1956). A Blood Group Survey in West Nakanai, New Britain. *A. J. Phys. Anthrop.* **14**: 275-286.

Spriggs, M. (1997). *The Island Melanesians.* Oxford: Blackwell Publishers Ltd.

Swindler, D. R. (1955). The Absence of the Sickle Cell Gene in Several Melanesian Societies and its Anthropological Significance. *Hum. Biol.* 27: 284-293.

Swindler, D. R. (1959). *A Racial Study of the West Nakanai of New Britain.* Ph. D. Dissertation, University of Pennsylvania. Available from Ann Arbor: University Microfilms International.

Swindler, D. R. (1962). *A Racial Study of the West Nakanai.* Philadelphia: The University Museum

Swindler, D. R. (1991). The Craniofacial Complex in the Lakalai of New Britain, Melanesia. *XVII Pacific Science Congress Publication* (eds.), T. Brown and S. Molnar, pp. 21-26.

Swindler, D. R and M. I. Weisler. (2000). Dental Size and Morphology of Precontact Marshall Islanders (Micronesia) Compared with Other Pacific Islanders. *Anthropological Science,* **108**: 261-282.

Swindler, D. R. (2005). A Review of Dental Morphological Traits in Oceania. In: *A Polymath Anthropologist, Essays in Honour of Ann Chowning,* (eds.), C. Gross, H. D. Lyons and D. A. Counts, Research In *Anthropology & Linguistics,* Monograph No. 6, The University of Auckland, pp, 35-44.

Valentine, C. A. (1958). *An Introduction to the History of Changing Ways of Life on the Island of New Britain*. Ph. D. Dissertation, University of Pennsylvania. Available from Ann Arbor: University Microfilms International.

Valentine, C. A. (1961). *Masks and Men in a Melanesian Society. The Valuka or Tubuan of the Lakalai of New Britain*. Lawrence: University of Kansas Publications.

Valentine, C. A. (1963). Men of Anger and Men of Shame: Lakalai ethnopsychology and its Implications for sociopsychological Theory. *Ethnology* 2: 441-477.

Valentine. C. A. (1965). The Lakalai of New Britain. In: *Gods, Ghosts and Men in Melanesia: Some Religions of Australian New Guinea and the New Hebrides* (eds.), Lawrence, P. and M. J. Meggit, Melbourne: Oxford University Press, pp. 162-197.

Valentine, C. A. and B. Valentine (1979). Nakanai: Villagers, Settlers, Workers and the Hoskins Oil Palm Project (West New Britain Province). In: *Going Through Changes: Villagers, Settlers and Development in Papua New Guinea*, (eds.), Valentine, C. A. and B. Valentine. Port Morseby: Institute of Papua New Guinea Studies, pp. 49-71.